Who Gave Pinta to the *Santa Maria?*

ALSO BY ROBERT S. DESOWITZ

The Malaria Capers

The Thorn in the Starfish

New Guinea Tapeworms and Jewish Grandmothers

Who Gave Pinta to the *Santa Maria?*

Torrid Diseases in a Temperate World

Robert S. Desowitz

W·W·NORTON & COMPANY

NEW YORK · LONDON

Printed in the United States of America
First Edition

The text of this book is composed in Caledonia
with the display set in Baskerville
Composition and manufacturing by The Maple-Vail Book Manufacturing Group

Library of Congress Cataloging-in-Publication Data
Desowitz, Robert S.
Who gave Pinta to the Santa Maria? : torrid diseases in a temperate world / by Robert S. Desowitz.
p. cm.
Includes index.
ISBN 0-393-04084-4
1. Tropical medicine—America. I. Title.
RC962.A45D47 1997
616.9′883—dc20 96-44741
CIP

W. W. Norton & Company, Inc., 500 Fifth Avenue, New York, N.Y. 10110
http://www.wwnorton.com

W. W. Norton & Company Ltd., 10 Coptic Street, London WC1A 1PU

1 2 3 4 5 6 7 8 9 0

For Carrolee

Contents

Preface

IN THE BEGINNING the intent was simple; this book would relate, concisely, how some of the classically "tropical" diseases came from the Old World to the New World. There would be a straight story line. But the book material took over, and the story weaves and dodges in faithful reflection of how things really are rather than how writers and their editors would like them to be. Over the aeons of human existence, infectious microbial-parasitic diseases appear, disappear, and shape the social, economic, and cultural fabric while the social, economic, and cultural fabric shapes the infectious diseases. What a mishmash! I think a wonderfully fascinating one. So I hope the reader will not only forbear but also take enjoyment as we weave and dodge through some of our medical history from 50,000 B.C. to 2500 A.D.

In writing this book I had a support group of old and new friends and colleagues, generous of their time and advice: Dr. Louis Miller, head of the Laboratory of Parasitic Diseases at the National Institute of Health, gave his customary sagacious advice and suggestions; Dr. Tom Monath of the Oravax Corporation kept me straight on yellow fever; Dr. Tom McCutcheon shared his new, exciting research findings on malaria with me and in doing so threw my nice orderly ideas on malaria in the pre-Columbian Americas into total confusion; Dr. Richard V. Lee, professor of pediatrics, obstetrics and gynecology, and geographical medicine at the State University of New York at Buffalo, School of Medicine, kept me straight on syphilis and shares a kindred sentiment on the nature of health and the value of yellow pickup trucks; and Profs. Marc Wéry and Domenique LeRay of the Prince Leopold Institute of Tropical Medicine, Antwerp, offered their hospitality and shared their great fund of knowledge during my stay there as a Damien Foundation–Fulbright Research Fellow. Mrs. Nancy Miller suggested the jacket cover design. My final words of appreciation are for my wife and best friend, Carrolee.

Honolulu, Hawaii
Antwerp, Belgium
Pinehurst, North Carolina

Introduction: Tropical Diseases—As Anglo-American as the Heart Attack

IN 1962, when Indonesia's dictator, Sukarno, was in flower, I attended a meeting of the Southeast Asian Ministers of Education Secretariat Tropical Medicine Project. The representative from Indonesia opened the session by hotly proclaiming that there was no such thing as "tropical" medicine; it was a colonial term of denigration (the implication being that the whites were hygienic and the natives unsanitary). The head of the project, Prof. Chamlong Harinasuta, the Thai physician and scholarly scientist, was equally adamant in his rebuttal. *Everyone* knew what tropical medicine was. Chamlong won the day: Tropical Medicine Project it was to remain, and still is. To suggest that tropical diseases were no more exotic than, say, a common cold or heart attack in Buffalo might well discourage the United States and other governments who were funding the project. Who was correct? Was Chamlong a pawn of the neocolonists or is there, in truth, a special segment of medical science geographically unique to the tropical regions? The answer is probably "yes"; both were correct in their interpretation of what constitutes a tropical disease.

Some diseases have always been confined to tropical locales beyond the temperate zones. Most, if not all, are infections caused by parasites, bacteria, or viruses. Many of these pathogens are in exquisite biological dependency to a spineless third party, such as a mosquito or a snail, for transmission from one host to another. But in truth, only a few exclusively tropical diseases come to mind. Most have ranged so widely as to be almost cosmopolitan. Malaria, that exemplar of the tropical disease? It once sickened and killed Californians, North Carolinians, Hollanders, and the English residents in the Thames estuary marshlands of London. Yellow fever? Killing epidemics raged from Boston to New Orleans between 1673 and 1905. There was even a mini-epidemic in Swansea, Wales in 1865. Cholera? There were fearful epidemics throughout pre-sewer North America. And epidemic cholera, the most feared disease of the nineteenth century, has again come to the Americas with 332,562 cases in Peru, 2,690 cases in Mexico, and 26 cases in the United States during 1991. Hookworm? It was once the great desanguinator of the rural U.S. southerner, the California miner, and the Alpine Swiss tunneler of St. Gothard. And what of the giant, elephantine, scrotum of filariasis which once afflicted the male citizen of Charleston?

Of those pathogens now confined to tropical countries most depend on their geographically limited hosts. The virus of Lassa fever, for example, normally circulates among African rodents, and from them, humans contract the infection. Lassa fever is a zoonosis—an infection from animal to human. Some pathogens are geographically confined because of the idiosyncratic behavior of the human hosts. The "whatever it is" pathogen of kuru is or was limited to the Fore people of Papua New Guinea who ritualistically ate the brains of their enemies.

Even so, the most "tropical" pathogens can break out from their motherland to the cool north and south as well as elsewhere. We are not isolated or insulated from the scourging microbial and parasitic diseases of the tropics. The Lassa fever virus brought in by an African stowaway rodent might, for example, adapt to the sewer rats of New York or London. Confronted with the most frighteningly deadly viruses yet to emerge from the tropics, the Ebola and Marburg viruses of Africa, there is much anxiety in the West that the doomsday bugs can spread from their "normal" habitat. Certainly, the past history of tropical temperate disease exchange can give us scant comfort. Our world has never been compartmentalized.

Who Gave Pinta to the *Santa Maria?*

Chapter 1

Coming to the Americas: 50,000 B.C. to 1492 A.D.—The Humans

MY GRANDPARENTS sailed to America from Austria and Galicia in the 1880s, swept up in the wave of nineteenth-century immigration. Neglectful of family history, I can't trace my origins back more than a century. Still, even that small fragment of history pleases me. We take comfort in knowing where we come from and the historical causes that ultimately brought us here. I mention this bit of autobiography because on occasion I visit my two young grandsons in Florida and grandfathers of all societies have a ritualistic obligation to (1) bring toys and (2) pass down the mysteries and genealogy of the tribe. The problem is, my grandsons have the privilege of being part American Indian, and there is still some uncertainty as to their deepest roots. There are some clues, some speculations. And that's how it should be; grandfather stories should never precisely cleave to facts.

It might also be that my grandchildren will eventually want to know what baggage of pathogens their ancestors brought with them from the Old World and what microbial dangers they faced in the New.

A speculative look at Amerindians and their diseases prior to 1492 presents scenes of shifting complexity. Let us first consider the weather.

Actually, there are two "weathers," inside weather and outside weather. Its always summer in our insides, 37°C

(96.8°F)—the humid tropical heat of the metabolically regulated healthy human body. Worms and germs flourish in that milieu. The problem for the worms and germs is that of going from host to host to perpetuate their species since they do not enjoy, as we mammals do, the uninterrupted heat of sex and gestation. The microbe's and the parasite's journey of perpetuation could expose them to the harsher climate and conditions of the outside world. Of course, the sexually transmitted pathogens have solved their perpetuation problem in a most rational way, and other microbes have developed other strategies such as forming resistant spore stages. But for many of the pathogens it can be bitterly cold and hostile on the outside. Thus, the weather in America at the times of successive human migrations would certainly have dominated disease epidemiology.

A (hypothetical) wormy, malarious Asian immigrant comes to America, circa 20,000 B.C. (or earlier, depending on which evidence you accept). He arrives in Alaska and he defecates. Frozen feces. Put an egg in the refrigerator and the embryo won't develop. That's true whether the egg is chicken or parasitic worm; all eggs need warmth to develop. The migration maven paleoparasitologists have used this basic biological fact in sleuthing the human colonization of the Americas. Their train of logic begins with the fact that humans in Africa, Asia, and Europe have been massively parasitized by intestinal worms for many thousands of years. The common great triad of gut worms are the roundworm *(Ascaris lumbricoides),* the whipworm *(Trichuris trichiura),* and the hookworms *(Ancylostoma duodenale, Necator americanus).* All three worms are strictly parasites of humans; animals can't become infected—no humans, no worms. So where did these Old World parasites come from when they parasitized the Amerinds?

All have a life cycle in which the eggs embryonate to the infective stage (or hatch in the case of hookworm) while in the soil. They require a minimum balmy temperature of 20°C (68°F) to 25°C (77°F) for them to do so. Only after they have embryonated, that is, incubated to contain larval worms, are the eggs infective, transmittable to a new human who will ingest these ova in contaminated soil or food. The hookworm is a little different; its larva develops rapidly within the egg and under optimum conditions hatches from the egg and then dwells in the soil. Along comes our barefoot pilgrim and the hookworm larva penetrates the skin and, after a complex migration through the body, comes to its home in the small intestine where it sucks blood assiduously from the vessels of the intestinal mucosa. If our hypothetical scenario holds true, all the original Amerind migrants would have been "cold sterilized" of their intestinal parasites by the time they moved south to the more parasite equable climes. Thus any pre-Columbian worms in Amerinds would signify that there were contacts and introductions before Columbus and his wormy crew reached our shores.

Another example is the mother of fevers, malaria. This disease, exquisitely dependent on temperature for its perpetuation by transmission through the mosquito, eventually became a major scourge of North and South Americans. Human malaria is a constellation of four protozoan parasites of the genus *Plasmodium,* all of which have an obligatory cycle of development in Anopheline mosquitoes and *only* anophelines, not the little brown nuisances (Culicines) that bite you at night or the spotted ones that bite you in the shade of late afternoon (Aedines). Those transmit some nasty viral diseases that devastated the Americas, but they don't transmit the malaria parasites. The malaria parasites require a certain minimum temperature, 20°C (68°F) to 24°C (75°F) depending on the species of the malaria, to undergo their complex cycle of transformation to the infective form in the anopheline. Many mosquito species can live in temperatures below the malaria parasite's life limits. Indeed some anophelines, as well as other mosquitoes, can winter over to await spring's warmth. However, an anopheline without malaria is just another damned nuisance.

Today's climate doesn't gauge yesteryear's weather. Throughout the hundreds of millions of years that the Earth has been a planet of the living, climate has been a sometime thing. The table on the following pages illustrates how climate has bounced around these last 400 million years and how it is expected to change again during the next century. Numerous causes are responsible for these climatic changes. For one thing, our Earth is not cemented in its heaven; it wobbles. Sometimes it is closer to the sun and sometimes it is farther away. Thus solar energy increases and decreases, possibly in a cyclical fashion every 100,000 years or so. Then there are natural and cataclysmic climate-altering events, like the impact of an asteroid. Nor is the greenhouse effect novel to our time. In the very ancient past, long periods of volcanism spewed massive amounts of carbon dioxide into the atmosphere. The industrial age has led to the well-publicized discharge of carbon dioxide, fluorocarbons, and other greenhouse, or ozone-depleting, gases. Meat and milk for the millions has led to the great expansion of the cattle industry. The collective flatulence of cows brings great amounts of methane into the air—more greenhouse. And so when we speak of migrations and pathogens, the climate of the times is always an influencing presence.

When and where: If we throw caution to the winds of hypotheses, the first Americans arrived 50,000 years ago, give or take 5,000 years. The more accepted scenario is that sometime around 20,000 B.C. they crossed from Siberia to Alaska, island hopping or land bridging by way of the Bering Straits. They were northern mongoloids with type O blood and the identifiable bits of genetic markers that characterize the Amerind or northern mongoloid. Theirs was a cold crossing during the last glacial age when the ice extended from the Polar caps. The Amerinds rapidly fanned out to the east and south; by 16,000 B.C. they had reached our East Coast and northern South America. A few thousand years later the Indians were everywhere, from Alaska to Tierra del Fuego. The latecomers were the Eskimos, the Inuits who arrived in Alaska from the Kamchatka peninsula about 10,000 years ago. Theirs was a lateral spread across the frozen North and forest tundra to Greenland. Still later, about 4,000 to 5,000 years ago a third immigration wave flowed from the Siberian forests. The immigrants came to the American Northwest Coast and stayed there to become the Tlingit, Athapaskan, Haida, and Eyak tribes. From the evidence of linguistic and genetic homologies it appears that one group wandered from the Northwest Coast to the Southwest. These were the Navajo. That is the textbook account of how humans first came to the Americas, although from textbook to textbook there is controversy over the details.[1]

So what were people doing in northeastern Brazil 50,000 years ago?

1. There is a curious "big bear" theory that has been proposed to account for the rapid dissemination of the Siberian migrants. According to this rather far-out theory, on crossing the Bering Straits, the migrants were met by a ferocious, enormous (now extinct) bear, the Giant Short-Faced Bear *(Arctodus simus).* It was long-limbed and so large that the modern Alaskan Kodiak would look like a puny pigmy beside it. It was also believed to be very short tempered; the expert on the animals of the Pleistocene period, Björn Kurtén, noted that it "may well have been the top predator of the world." Perhaps those first immigrants to America, on meeting the Short-Faced bear, moved rapidly to a less hostile neighborhood to the south.

CLIMATE THROUGH THE AGES

THE TIME (YEARS AGO)	THE WEATHER	WHAT'S HAPPENING
400 million	Cold everywhere	Continents amalgamated into Gondwandaland. An ice age. Ice in the Congo basin reaches what will be Brazil.
60 million	Warm and sunny everywhere	Subtropics in present temperate zones. No polar icecaps. The Rock mountain area is an ocean. Dinosaurs die and mammals begin their major evolution.
50–30 million	Warm and sunny; tropical hot in California; temperate in Alaska	Monkeys in Africa split into two types. Platyrrhine type comes to South America and becomes extinct in Old World.
25 million	Cool	Ice age begins but climate still mild.
5 million	Cold	
3–2 million	Getting colder	Continents in their present position. Extensive glaciation in North America. Isthmus of Panama raised from the sea allows faunal interchange between North and South America, but no monkeys come to North America. Temperate in Antartica.
1 million–present	Cool and variable	Cyclical retreats and advances of glaciers.
800 thousand	Cool in the North, warm in the midlatitudes	
400 thousand	An ice age of glaciation in northern North America	Glaciation creates a 600-mile-wide land bridge between Alaska and Siberia. Bovids, such as bison, cross over from Asia to America.
50–35 thousand	Cool to cold in northern North America; hot and humid in the South	The last glaciation begins. Sea levels low. Amerind-type humans come to America? Camels, mammoths, cheetahs, sloths, zebras, and horses in Las Vegas and other parts of the West.

THE TIME (YEARS AGO)	THE WEATHER	WHAT'S HAPPENING
18 thousand	Cool	Last glacial maximum in North America. Sea levels 200 feet lower than present. Amerinds to America.
12 thousand	Cool winters and hot summers	Last deglaciation. Extinctions of many large mammal species in North America. Earth farther from sun in January but closer in June than now. Lakes in the Sahara desert.
10 thousand	Temperature about like now	Inuits come to America. Amerind settlement complete throughout the Americas. Some Amerinds begin agriculture.
7 thousand	Cool and pleasant	Northern ice sheets melted. Postglacial spread of hardwood forests into northern Canada
5 thousand	Cool to warm	Warmest postglacial time. Deserts moister.
4 thousand	Cool and getting colder in the North	Forests in northern Canada recede and pines replace the oaks.
900	No rain in sight	A long drought in the American plains and prairies. Agriculture collapses; many Indians move south.
450, 150 (1550 A.D.; 1850–1990 A.D.)	Cool and cold	Little ice ages in America; glaciers advancing.
100	Warming	Industrial age begins; fossil fuel usage 60 times greater than preceding century. Droughts in the Great Plains during 1890s and 1930s.
Now and 100 years from now	More warming	More fossil fuel usage; more greenhouse effect. Global temperature expected to rise 3°C by 2090 as atmospheric CO_2 is doubled.

In a remote region of northeastern Brazil, a landscape of high sandstone cliffs and scrub bush, caves, and rock shelters are adorned with spectacular Lascaux-like paintings. It had been assumed, for many years, that this was the art of an Amerind tribe who lived there 7,000 to 8,000 years ago. In 1986 a French anthropologist, Dr. Niède Guidon, upset that applecart, and virtually every other specialist in New World prehistory, by declaring that the oldest paintings dated to 32,000 B.C. She had the laboratory evidence, carbon 14 dating of charcoal from domestic fires, to back up her claim!

In 1993 a doctoral candidate who had been "dissertation digging" at Pedra Furada since 1984 sat in Paris before his committee to defend his thesis that human habitation went back *not to 32,000 B.C. but to 50,000 years B.C. or more.* That committee had reviewed the monster four-volume, 15-pound thesis and sat for four hours to hear the student paleoanthropologist, Fabrio Parenti, make his defense. Parenti argued that he had found assemblages of quartz pebbles at the 50,000 years B.C. sediment stratum in front of the rock shelter that were not randomly dispersed in natural fashion but were arranged in the collected fashion of humans.

Parenti's jury accepted his thesis, but for others the jury is still out. The sceptics have a problem in accepting the pebble proof of human habitation in 50,000 years B.C. Brazil. Then too, there is the problem of discontinuity. There are no discovered signs of human habitation that early anywhere in North America. Thus, those early putative Brazilians conflict with the historical orthodoxy that the human occupation of the Americas was by a southward expansion from the Bering Strait "beachhead."

One obvious catch to all this is the implication that if there were 50,000-year-old Brazilian Amerinds who came from Bering Strait transmigrations, then there would have to be 50,000+-year-old Siberians and no one knows whether humans had colonized Siberia that early or earlier.

The first human occupation of Siberia has been thought to have been between 20,000 and 30,000 years ago—a time which didn't jibe with the Pedra Furada advocates. That notion persisted unchallenged until a Russian husband and wife team of archaeologists, Yuri A. Mochanov and Svetlana Fedoseena, found a collection of chipped rocks at a site along Siberia's Lena River. To the untutored eye they didn't look like much of anything more than broken rocks. But the Mochanovs maintained that they were of human manufacture, simple tools made by smashing one rock against another. And what's more, the Mochanovs dated them at 3,000,000 years B.C.! New dating analyses by a technique called thermoluminescence date the artifacts at 500,000 years B.C. That is still very, very old—a time when our immediate ancestor, *Homo erectus,* is believed to have dispersed from Africa. And if true, as Dennis Stanford of the Smithsonian Institute has noted, ". . . if people were dealing with the cold that far north in Siberia 500,000 years ago, then a little bitty ice age like the Wisconsin (the name given to the ice age of that period) isn't going to stop you from getting to America." Thus if we can accept the specious factor, the Mochanov assertion, then the Pedra Furadan as a Siberian descendent becomes a logical possibility.

Chapter 2

Coming to the Americas: 50,000 B.C. to 1492 A.D.— The Worms and Germs

THE (CHARRED) STICKS and stones of Pedra Furada were not bones, quite the best proof of human presence. There is, however, another human remnant—feces. Under favorable aridity, feces will desiccate to a stone-like consistency and become coprolites—an archaeological term for fossilized stools. Humans without indoor plumbing (like many animals) often have customary places, not far from their habitation, where they relieve themselves. The coprolites can be sectioned to permit microscopical examination of their contents. Food remnants, the eggs of parasitic intestinal parasites, and the resistant cystic stage of intestinal parasitic protozoa (such as that of the diarrhea-causing *Giardia lamblia*) will also be preserved over the millennia. Scientists are beginning to apply the powerful, probing modern immunogenetic technology to the examination of coprolites, as well as to other organic relics of ancient life. In the near future, scientists will surely identify the bacterial and viral infections of our ancestors by mapping their DNA "footprints" in the fossilized feces.

But even with relatively unsophisticated, nonmolecular microscopical techniques a picture of ancient life has been reconstructed from clues in the coprolites. For example, from the communal privies used by the citizens of tenth-century Winchester and fifteenth-century Worcester, England, large numbers of preserved roundworm and whipworm eggs were found. Those people were heavily parasitized—attesting to the miserable state of British medieval sanitation. From seeds and pollen in the coprolites we know what herbals these people took for what ailed them—the black nightshade, the deadly nightshade, hepbane, self-heal, and the water dropwort. Seven hundred years later, in the seventeenth century, the English still had parasites galore. The great physician-physiologist of that time, William Harvey, departing from his seminal work on the circulation of the blood, spoke of the more mundane intestinal conditions of his fellow English-people by declaring, "Worms in all the guts, nothing so common as worms."

From the parasites recovered in the coprolites and the few American mummies that have been found, we can attempt to construct a likely scenario. The premise: any parasitic worm (or its telltale egg) whose "ancestral home" is Africa or Asia found in American human specimens dated as pre-Columbian indicates that there was human contact between Asia or Africa before 1492.

This was the approach of the Brazilian scientists at Pedra Furado. And feces they found—hard as a rock—at the 30,000 B.C. level of sediment. In these coprolites were the unmistakable barrel-shaped, plugged eggs of the whipworm, *Trichuris.* But this was not proof positive that they were from a parasite of humans, and a more searching examination showed how easy it was to be led down the garden of hypothesis's path. There are several species of *Trichuris* other than that of the human whipworm, *Trichuris trichiura,* and the eggs of all of them look pretty much alike under the microscope. Moreover, closer examination showed that the Pedra

Furada coprolites didn't look "human," that they were kind of kidney shaped in form. The Brazilian paleoarcheologists came to the disappointing conclusion that these were the droppings of a caviid (guinea pig-like) rodent, *Kerodon ruprestris,* that had migrated to South America at the time of the great continental breakup of 40 to 25 million years B.C. The whipworm was a parasite of this animal. *Kerodon ruprestris* still exists in the Pedra Furada area, but when the scientists trapped some of these animals, they found, curiously, that they no longer were parasitized with the whipworm as they had been 30,000 years before. So the worm (eggs) looked the same but the feces were different, and the identity of these humans (if indeed they were present) and where they came from remain a mystery.[2]

Was there truly an ancient "cold screen"? Were the Russian-Siberian immigrants to Alaska free of what we call soil-transmitted intestinal parasitic worms?[3] Thus, the pinworm

2. In this case the parasite eggs looked "human" but the feces looked "animal." In epidemiological sleuthing the opposite can also occur. Some years ago, large numbers of very human looking turds began to appear on the sandy, pristine shores of Kualoa Beach Park in Oahu, Hawaii. The health authorities were very upset because this might be a health hazard and, even worse, a threat to the tourist industry. Sharks are bad enough but feces would be intolerable. The law and political authorities were concerned because they didn't know who was defiling the beach—random vandals, the homeless, or Hawaiian nationalist-royalists making a political statement, so to speak. Examination of these feces revealed that there were numerous protozoan parasites and worm eggs of many types but none were of species parasitic in humans. They were parasites of sea turtles, and these were the "vandals" coming to the beach at night for purposes best known to turtles, in the course of which there was some indiscriminate defecation.

3. The parasites are called soil transmitted because the usual way of transmission from person to person is in fecally infected soil containing the eggs or infective, skin-penetrating larvae. Tourists to the wormy regions of the world may get theirs via the locally grown green salad. There's nothing quite like feces for organic gardening. The other intestinal parasite, the pinworm *(Enterobius vermicularis),* that itching scourge of the collective middle-class anus, doesn't require soil for transmission. Pinworm eggs

could have come over with the original migrants. However, there are no 20,000-year Siberian mummies to verify our assumption. All we have are the modern Russians. And a wormy nation they are. Infection rates, in fact, were so high in Lenin's time that the regime issued the manifesto "Devastate the Worms." This official concern prompted extensive surveys of intestinal parasites to be undertaken throughout the Soviet Union. They showed that the aboriginal peoples of Asiatic Russia, living above 60° north latitude, are free of the two most common intestinal worms, the roundworm *(Ascaris)* and the whipworm *(Trichuris).* A little bit farther south of Siberia, in the Far Eastern zone, the worms begin. But even there, the warm-loving hookworms are absent.

Few pre-Columbian desiccated Indians and/or their fecal specimens exist, but those few support the notion of the clean Siberian coming to America. The ancient North American human mummies and coprolites have come from sites with romantic-sounding names such as Dust Devil Cave (Utah, 6800–4800 B.C.), Big Bone Cave (Tennessee, 500 B.C.), Turkey Pen Cave (Colorado, 700 B.C.), Antelope House (Arizona, 1300 A.D.), and Salmon Ruin (New Mexico, 1300 A.D.). These and other pre-Columbian sites in the Aleutians and the Northwest Coast have, so far, yielded nothing that indicates a soil transmitted parasite. Our old anal nemesis, the pinworm, was there but that's a hand-to-mouth traveler.

The pre-Columbian South and Central American Amerinds should have been as free of intestinal parasites as their cousins in the North. They should have, in particular, been free of hookworms. Hookworm parasites are transmitted by infective, skin-burrowing larvae in the soil. These tempera-

become infective, embryonated, very rapidly and stick to fingers, bedding, and such. Hand-to-mouth transmission is the usual mode of infection with pinworms.

ture-sensitive larvae would have been quickly frozen had they come across with their human hosts in the Bering Strait migrations.

Thus, hookworms would leave the best footprint of pre-Columbian contacts. Their presence in coprolites and mummies would speak both of contact(s) and where those ancient travelers hailed from. This is because there are two kinds of hookworms in humans and their ancestral "homes" are in different parts of the world. *Necator americanus,* whose name must have been bestowed by a taxonomist with a warped sense of either patriotism or geography, comes from Africa. The other hookworm, *Ancylostoma duodenale,* comes from Asia.

Unlike the North American continent, between the remote, clouded time of 50,000 B.C. and 1492 A.D. the fossil worm record reveals pre-Columbian visitors to the American Southern Hemisphere.

The Brazilian paleoparasitologist L. F. Ferreira and his colleagues of the National School of Public Health in Rio de Janeiro continued to investigate in the Pedra Furada area. At one site they struck metaphoric gold, a treasure trove of human coprolites. Those fossilized stools had been passed 7,320 years ago and they contained hookworm eggs! This was indeed exciting, but the species of hookworm producing those eggs could not be determined. The eggs of both species of hookworm are morphologically identical; one needs the adult worm for those morphological landmarks that distinguish one species from another. Also, it would have been comforting to find at least one more site bearing the hookworm sign of transoceanic pre-Columbian contact.

Confirmation came from a dry mummy high in the Peruvian Andes, a man who had died some 2,800 years ago. In the mummy's intestinal tissue there were helminth parasites that when examined by scanning electron microscopy, were deemed to be hookworms—*Ancylostoma duodenale* hookworms from tropical-temperate Asia!

But what Asians could have made that seemingly impossible journey so many thousands of years ago? The archaeologists Betty J. Meggars and Clifford Evans believe that 5,000 years ago Japanese from Kyushu were swept by misadventure to Ecuador's shores, bringing with them pottery and, as some parasitologists contend, the hookworm.

In 1961 Meggars and Evans and their Ecuadorian colleagues were working a dig in Valdivia, a seaside town on Ecuador's Pacific Coast. Valdivia has been a site of human occupation for over 5,000 years. Layer on layer of relics give testimony to successive inhabitants. At the lowest stratum, representing a period of neolithic toolmakers, there were mollusk shells, simple tools of bone and stone, *and* pottery of surprising design and sophistication. The pots which "suddenly appeared" were red slipped, incised to make a distinctive decorative pattern, and rimmed by a rising series of ridged castellations. The pots were in fact identical to the pottery of Japan's southernmost island, Kyushu, during the middle Jomon period of 3,000 years B.C.—the same time that carbon 14 dating had given for the age of the Valdivian pots.

The Kyushu people of the middle Jomon period were gatherers of the tidal flats and fishers of the deeper offshore waters. Meggars and Evans believe that a canoe or canoes swept too far out to sea by inclement weather would have been carried by the strong west-to-east current whose path leads to the Ecuadoran coast. It would have been a grueling voyage, but the hardy neolithic Japanese are thought capable of surviving it. It would also have been a one-way trip; there is no alternating current back to Japan. The Kyushans might well have been accepted by the Valdivians who shared a simple neolithic fishing culture. Thus, the assimilated Japanese would have introduced their pottery and pottery-making technique. But would they also have introduced their parasites? Did they have parasites? Here again we must extrapolate, as we did for the Siberians.

We would not expect that the technologically sophisticated Japanese to have had a Third World level of parasitism. But that was the case of Japan until the mid-1950s. In 1951, 50 percent of those living in Kyushu had hookworms, some so serious as to be fatal. In 1951 alone, hookworm killed 1,261 Japanese. Indeed, until the late 1950s when control programs were initiated, Japan was a parasitologist's paradise. Consider: Every Japanese medical school has a department of medical parasitology. The most pathogenic blood-inhabiting parasitic worm is named *Schistosoma japonicum.* So, it is for good reason that I have my personal tradition of introducing each year's medical school lecture on filariasis (elephantiasis) by projecting a slide of a wood-block print. The print made by Hokusai, that superb nineteenth-century Japanese artist, shows a Japanese gentleman with his enormous scrotum in a sling. And one of the most neurologically damaging mosquito-transmitted viruses is called the Japanese encephalitis virus. These parasitic and microbial infections had been entrenched in Japan for many, probably thousands, of years. This would fit our hypothesis that both hookworm and pots attest to pre-Columbian contact from Japan.

Then there were the transcontinental infections that followed the first migrants to America

When I lived in Honolulu, every mid-April I told my plover that he was an idiot. Why should he fly, nonstop, 4,000 to 5,000 miles simply for sex. "Stay here," I'd tell him. "There are lots of great chicks in Honolulu. Maybe I can fix you up with that nice Hawaiian stilt I know." But my plover rejected all blandishments. In his new, elegant courting plumage of jet black bib and white shawl, he would, on a late-spring morning, leave his feeding territory of that small patch of grass fronting my condominium's tennis court and wing his way north to Alaska or Siberia. In September, *Deo volante,* he'd be back bug picking on the Hawaiian turf.

In the very far North during the short, warm, humid summer our plover rendezvoused not only with his ladylove but also with the mosquitoes, midges, ticks, and biting flies that awaited his return.

During my many years in sultry climates I have been bitten by tsetse flies, sand flies, blackflies, horseflies, and mosquitoes of many varieties as well as by ticks and mites; but nowhere in the tropics have I been attacked by insects so viciously and massively as when flogging trout streams in the Arctic watershed. The abundance of blood-sucking insects that proliferate during the Arctic's short summer is simply amazing. They make life hell for humans and wildlife. They also transmit viruses that can affect human health. Arthropod-borne viruses are known collectively as arboviruses.

The "natural" hosts of many arboviruses are migratory birds, such as the waterfowl that make epic seasonal flights to their Arctic breeding grounds. Most arboviruses are not too fastidious and can spill over into the mammalian populations—wild, domestic, and human. Some arboviruses are transmitted by ticks (arthropods more related to spiders than insects). Ticks and mites infest the nests of the waterfowl where they have easy access to a blood supply of chicks, which they infect with virus as they feed. Moreover, some viruses can be perpetuated from tick to tick via the tick's egg. If a female tick carries that virus, it can enter her eggs and lie dormant over the winter. The egg develops; the larval tick hatches already loaded with virus and capable of passing it on to the bird or mammal on which it takes its first blood meals. From the birds without frontiers and the mammals without frontiers and the swarms of vector mosquitoes and ticks at all frontiers, trans-Arctic zoonotic arboviruses probably posed a health threat to the first Americans when they colonized Alaska. They may also have carried viruses in their person or in their accompanying domestic animals that became forever endemic in the Americas. Indeed some of those viruses may have proved highly lethal to the immunologically naive American wild mammals. Shortly after the arrival of humans in North America, many of the large mammals became extinct. Did a human virus or an introduced zoonotic virus from a domestic animal kill them off?

The historical epidemiology of viral diseases is a vexing area of research because viruses are so small and often so mutable. "Here today, gone tomorrow," as well as, "Gone today, here tomorrow," characterizes the changeability of viruses and how they come and go as threats to humans. Lassa fever and, most notably, AIDS are two diseases that came "from out of the blue" in our time. A great disease of the past seemingly without parallel in our present time is the Plague of Athens. That mysterious disease killed one-third of the Athenian population over a 20-year period beginning in 430 B.C. It was a new and unfamiliar plague, swift and killing, striking "like a wolf on the fold" in epidemic proportion. The only clinical description we have is from the philosopher-general Thucydides who was himself stricken but recovered. It was a highly contagious disease giving rise to high fever; respiratory, intestinal, and neurological involvement; a rash; and gangrene of toes, fingers, and penis.

Nothing quite matches the Athenian plague in the present clinical logbook or in the subsequent historical medical logbooks. It disappeared with no evident recurrence elsewhere. It has, for many years, challenged diagnosticians who have proposed some 30 causative, mostly viral, agents. I added a thirty-first and speculated that it was a form of Lassa fever (Desowitz, *MD Magazine,* May 1994). Lassa mostly fits Thucydides's account except that Lassa doesn't make your penis fall off; but why quibble over small details?

Then there were the native parasites and microbes of the Americas

When I began to think about the microbes and parasites that were present in the Americas prior to human occupation, I fell into the "only man is vile" trap. I should have known better. The romantic western hemisphere of my brain envisaged a prehuman America as a kind of peaceable kingdom, a pathogen-free land or at least a new world free of human-threatening pathogens. My saner self recognized that this was a crazy, sentimental notion. All living creatures have microbes. Microbes have microbes.

When the world broke up those many millions of years ago, the animals that came to be in the New World undoubtedly had within them a collection of microbes and parasites. In time, isolation from the Old World made these animals evolve into uniquely American species. And as they evolved, so did their viruses, bacteria, parasitic protozoa, and worms also become uniquely American species.

The rich variety of arboviruses peculiar to North America and the tropical Americas is impressive in a frightening way. There is, for example, WEE (western equine encephalitis), EEE (eastern equine encephalitis), and SLE (St. Louis encephalitis), mosquito-transmitted viruses ranging from northern Canada to Argentina in a massive natural reservoir of infection in reptiles, wild birds, wild mammals, and (now) horses. The virulence of these arboviruses varies from place to place, time to time, and person to person. Arbovirus disease ranges from the asymptomatic to the simple febrile headache to the fatal neurological meningioencephalitis. And not only humans and horses are affected; in 1984 7 of 39 whooping cranes in a captive breeding program were killed by the eastern equine encephalitis virus.

This great reservoir of infection was being silently transacted between wild animal and mosquito many thousands of years ago. As the early Amerindians descended into the mosquito latitudes, they would have encountered one of the encephalitis arboviruses, probably the WEE virus, and it would have gone to their brains. There is a broad range of disease caused by this virus, varying from individual to individual, from simple fever-headache-joint pains to the fatal disease of the nervous system in which there is a progression from drowziness to lethargy to convulsions to coma to death.

The Amerinds, like today's tourists, would have been attacked by new mosquitoes carrying new zoonotic viruses when they came to Central and South America. These viruses with strange-sounding names hint of far-off places—Mayaro virus, Maituba virus, Tacaribe virus, Machupo virus, Tamiami virus; they are relatively mild pathogens that "only" cause headache, fever, and intense joint and muscle pain. A more serious threat would have been the Ilheus and Venezuelan equine encephalitis arboviruses which can cause fatal brain damage. Here again, we do not have proof positive of the extent, if any, that these arboviruses infected the pre-Columbian Amerinds or what impact they had on their health. The modern DNA microbe hunters are beginning to turn their attention to these questions of medical history, and we are beginning to understand how the early human inhabitants fared in the American viral jungle.

Among the hostile indigenous American pathogens were members of the Hemoflagellate family

The Hemoflagellates affecting humans are divided into three kinds. First, there is the African sleeping sickness family of trypanosomes transmitted by the tsetse fly. The pathogen swims in the blood and has two clans, one in West and Central Africa, the other in East Africa. They are out of our pre-Columbian American consideration so we will leave it at that.

Second, and more pertinent to our story, are the leishmania. Many species of vertebrates including humans have their species of leishmania parasite. I once suggested, not too seriously (*The Malaria Capers,* New York: Norton, 1993), that a leishmania of lizards may have killed the biggest lizards of all, that the dinosaurs may have been exterminated by a leishmania parasite rather than a meteor. There are many leishmania species pathogenic to humans that are spread from the Mediterranean through Africa to the Indian subcontinent and onward to the tropical and not-so-tropical Americas. Although outside the time and place of this chapter, it is worthwhile to note that the leishmania of the Middle East have been of recent threat to American military fighting in Desert Storm. There have been about 40 cases diagnosed from that military operation. Iraq has been, for many, many years, an intense focus of endemic leishmaniasis of both types, the skin invading and the deep organ invading.

All leishmania travel via a minute biting midge, the sand fly. Of all the Hemoflagellate parasites, the leishmania have made the strangest and boldest of adaptations in life style. In their vertebrate-dwelling phase, they have become restricted to living within the macrophages, the frontline soldier cells of the immune system that have the function of ingesting and destroying foreign bodies and pathogens such as the leishmania. To survive and flourish within the macrophage "garbage disposal" cells of their vertebrate hosts, the leishmania possess a variety of defensive adaptations.

Although under the microscope all leishmania look alike, it is now known from the elegant, exquisitely specific DNA homology techniques that there are many, many species. About 30 species, or types, are peculiar to the Americas. Why so many American leishmania? It is believed that when Gondwanaland began to break up about 250 million years ago and the Western Hemisphere continents began the slow ocean voyage to their present geographical locations, they carried, like some enormous ark, the animals and plants that evolved to become the American fauna and flora. Those animals carried parasites and other microbes, and those too diversified, under conditions of continental isolation like so many Darwin finches (which now have been shown to have diversified very rapidly into new distinct types).

One group of leishmania became "dermatologists," specialists adapting to an exclusive life in the macrophages of the skin tissue. Here they cause an inflammation (which calls for still more macrophages to migrate to the invaded area providing still more "meat" for the parasites) leading to ulceration. Fortunately there is a vigorous immune response to most (but not, unfortunately, all) of these skin-invading leishmania, and the ulcers, in time, heal naturally without the need of drug therapy.

After the ulcers heal, there seems to be a lifelong immunity to reinfection. But the ulcers are, nevertheless, not always all that benign; they are disfiguring and some species or strains of American leishmania evade or depress the immune system to cause extensive, nonhealing lesions.

Where early Amerindians first encountered sand flies harboring the leishmania is not known. Today, the northern boundary is southern Texas where eight cases of dermal leishmaniasis have been recognized and recorded. The Texan sand-fly vector, that minute midge, has the sinister name of *Lutzimya diabolica.* The real kingdom of the American leishmania is to the south, a vast region from the Yucatan to northern Argentina.

The ecological niches, the "landscape epidemiology" of leishmaniasis within this vast region, are diverse, from the high, dry, cool Andes to, especially, the steamy rain forest. Here natural infectious transactions take place in the trees where the arboreal animals—the opposums, two- and three-toed sloths, and rodents—harbor the sand-fly-transmitted parasite in benign fashion. Humans are intruders in the forest, and the leishmania parasite is one of nature's no-trespassing signs. Thus when the early Amerindians colonized the rain forest, they would have been at risk. Even today, there is the skin ulcer that is the price extolled on the natives collecting the gum of the wild chicle tree in the tropical forest to provide you with the pleasure of chewing gum. These gum collectors, known as *chicleros*, consider the leishmanial lesions so common an occupational hazard that they call it the *chiclero*'s ulcer.

There is, however, a terribly malevolent leishmania of tropical America whose infection in humans does not terminate with a scar of self-cure. This species, *Leishmania braziliensis,* causes the mutilating disease espundia, which would have been a serious threat to the health of the pre-Columbian inhabitants of the neotropical forests.

In the dark ages of medical education when we actually lectured to students, it was microbiology's lot to have our classes scheduled immediately after the lunch break. We were thus confronted by somnolent, Pickwickian students whose blood had retreated, for digestive purposes, from the brain to the deep viscera. One effective revival technique was to project color slides showing grotesqueries of tropical disease. The enormous scrotal enlargement of elephantiasis (caused by a filarial worm parasite) was always good for a waker-upper, but the dictates of the new social or academic order required a more gender-neutral monstrosity—espundia. Espundia begins as a nonthreatening ulcerlike lesion of the skin typical of cutaneous leishmaniasis. And like typical cutaneous leishmaniasis the ulcer heals without chemotherapeutic treatment.

But unlike the typical cutaneous leishmaniasis in which the healing process is complete and even leads to a solid immunity against reinfection, espundia's *Leishmania braziliensis* leaves the skin to metastasize and invade the macrophages of the nose, and mouth or pharynx. It causes horrible destruction of the nasopharyngeal area of the face. The nose and mouth parts literally rot away. It is, today, one of the most feared diseases of tropical America especially so because it is so very difficult to treat chemotherapeutically with the onset of facial involvement. The interval between ulcer and facial involvement may be many months, even many years, and the victim usually makes no connection between the long-forgotten healed skin ulcer and the infection that is now ravaging the tissues of mouth and nose.

The problem for today's physician working in an espundia endemic area is how to treat a patient with a leishmanial skin lesion. Is it a self-curing type that doesn't require aggressive treatment? Is it *Leishmania braziliensis* that can lead to espundia years later? The lesions look alike; the leishmania recovered from the lesion look alike. Treatment for *Leishmania braziliensis* is most effective when it is aggressively given at the primary lesion stage, but it is long, costly, and unpleasantly toxic. Researchers can by sophisticated techniques, such as karyosome analyses, distinguish *Leishmania braziliensis* from the less virulent skin-invading species. But the techniques require considerable expertise and a sophisticated laboratory and are costly. Like so many remarkably effective diagnostic and therapeutic methods gifted by modern science, they are unavailable to the poor who really need them. It's the old, old question—who is going to pay? Certainly not the impoverished farmer of South America.

Were the early Amerinds at risk to this terrible zoonosis in the forest? The circumstantial evidence for the presence of muco-cutaneous leishmaniasis, espundia, in pre-Columbian times is found in the pottery of a the Peruvian pre-Inca Mochica people. Their terra cotta figures are of heads with extensive lesions of the mouth and nose so reminiscent of espundia. Also, the archaeological sites of the Mochica pottery is in the hot, humid region of Peru where muco-cutaneous is now (still?) highly endemic. That erudite Swiss medical historian–physician-pathologist–parasitologist–linguist, the late R. Hoeppli, considered the anthropomorphic Mochican pottery as depicting people stricken with espundia. But here we find an example of how widely and wildly experts can part. Another, equally distinguished scholar, the late H. Dietschy, looked at the same pots and concluded that the facial mutilations didn't represent the ravages of a parasite but were a tribute to the potato demon. *The potato demon?* Well, the potato originally came from Peru and the pre-Incans so loved the potato that they assigned a god-demon to it. According to Dietschy, the mutilations were made on the pottery faces (and, maybe, on human sacrificial victims) to give humans a potato-like face as a tribute to the demon. And the potato demon's name was Papamama.

When the Amerinds reached southern Mexico, they would have met what I consider to be the most terrible disease of the Western Hemisphere. The title of the opening chapter, "Tropical Diseases—As American as the Heart Attack," was intended to introduce the notion that North Americans have been beset since European colonization by diseases considered to be exotically tropical. That is a gringo narrowness; in the southern part of the Americas, from Mexico to Argentina, the main avenue to the heart attack, a major cause of sudden death, *is* a tropical disease—Chagas' disease—caused by a parasite of pure American ancestry. Today, within that vast region of entrenched endemicity it is estimated that over 60 million people are exposed to Chagas' and at least 18 million are infected. It is an infection lacking adequate chemotherapeutic cure or protective vaccine and it is coming north; from Texas to Detroit there may be as many as 100,000 cases.

The nineteenth-century South American physicians were confused by a variety of seemingly unconnected common illnesses—some mild, some acute, some chronic—and all of unknown causes. There was an ophthalmic condition in which a puffy swelling appeared suddenly around the eye. This occurred mostly in children but, because of its transient nature, was not considered serious. There was a fatal sickness of children that began with a fever and headache and ended with convulsions and coma. There was a chronic illness of adults and children characterized by prolonged low-grade fever, aching muscles, loss of appetite, and a generalized feeling of irritable "crappiness." And then there was a lot of heart disease, mostly in the middle aged—the irregular fast arrhythmic beat of tachycardia, the shortness of breath of early heart failure, and, finally, the massive fatal heart attack. Many of those who died in this way were found at autopsy to have grossly enlarged hearts. But the strangest and most frightening disease was one of the intestinal tract, a mega disease in which the colon and/or esophagus seemed to have simply lost power, stopped working, and become enormously distended, flaccid tubes. With digestion impaired and peristalsic waves stopped, the food mass failed to be processed and pushed to its anal conclusion. Feces accumulated within. Nutrition failed and the mega victim died. In 1910 these highly disparate conditions began to be threaded together as a common cause.

By the second decade of this century, parasitology's house was pretty much in order. The causative organisms of the major parasitic disease of humans were identified as well as their modes of transmission. Within this orderly taxonomic edifice, the pathogenic Hemoflagellate group of protozoans were considered to have a neat two-part delineation. The tsetse-transmitted trypanosomes swam freely in the blood whereas the sand-fly-transmitted leishmania were immobilized within the macrophages. To each his own. Or so it was thought until 1910 to 1920 when a Brazilian field researcher in Minas Geraes, his institute Director in Rio de Janeiro, and a Parisian parasitologist of the Pasteur Institute elucidated the Hemoflagellate of the third way—the causative organism of what became known as American trypanosomiasis or more commonly Chagas' disease.

He died in 1917 at 45, but during his brief life Oswaldo Cruz was Brazil's foremost medical scientist–public health authority. Cruz first came to prominence from his effort to control the yellow fever epidemic in Rio de Janeiro which killed some 15,000 people between 1891 and 1894. A grateful government bestowed a research institute on him (actually it was an old serum therapy factory-like laboratory) that was to become the still-famed Instituto Oswaldo Cruz. Cruz was a fine scientist, and unlike too many scientists-turned-administrators, he had the gift to select talented, innovative people for his staff. One of the Chosen was Carlos Chagas.

In 1910 a railroad was being built in Minas Geraes, a remote region of Brazil. Malaria was taking a high toll of the workers and bringing construction to a near standstill. Cruz dispatched Chagas, then 40 years old, to investigate the problem and apply whatever antimalarial measures might be feasible. Chagas was a scientist-physician of his time, a naturalist who believed that the understanding of human biology (and pathology) could be achieved only though the understanding of the biology of all living things. This spirit of epidemiological pantheism led Chagas to divert his attention from malaria and the anopheline to a bug that was sucking the blood from the peasants. The locals complained to Chagas of an insect, about an inch long, that by day lived in the cracks of the mud walls of their hovels and by night crept out to feed on them as they slept. They had several names for this arthropod vampire: *barbeiro* (the barber) and kissing bug because of its predilection to feed from the face and lips, and assassin bug because of the stealth of its attack. It also had the disgusting habit, later found to be of crucial importance in the transmission cycle, of taking blood at its head end and simultaneously defecating on its host from its hind end.

Charles Darwin, who became a blood meal while in Argentina, gives a graphic description of this reduviid (triatomid) bug in his book *Naturalist's Voyage.*

> At night I experienced an attack (for it deserves no less a name) of the *Benchuca,* a species of Reduvius, the great black bug of the Pampas. It is most disgusting to feel soft, wingless insects, about an

inch long crawling over one's body. Before sucking they are quite thin, but afterwards they become round and bloated with blood, and in this state are easily crushed. One which I caught at Iquiqui (for they are found in Chile and Peru) was very empty. When placed on a table, and though surrounded by people, if a finger was presented, the bold insect would immediately protrude its sucker, make a charge and, if allowed, draw blood. No pain was caused by the wound. It was curious to watch its body during the act of sucking, as in less than ten minutes it went from being as flat as a wafer to a globular form. This one feast, for which the Benchuca was indebted to one of the officers, kept it fat during four whole months; but after the first fortnight, it was quite ready to have another suck.

The bug may also have been the death of Darwin. Naturalists, despite the pleasures of their calling, have their trials with nature, particularly as they endure the attention of biting, stinging, blood-sucking, disease-transmitting insects and ticks. Darwin was no exception; during the five years from 1831 to 1836, on the voyage of the *Beagle,* he was repeatedly exposed to the kissing bug and thus at risk to Chagas' disease. He came home to England from his monumental voyage and was sick the rest of his life. Several medical historian–physicians have done a theoretical postmortem on Darwin and have concluded that his chronic lifelong illness—fatigue, irritability, repeated vomiting, flatulence—as well as his final illness and death from heart disease, was caused by a *Trypanosoma cruzi* infection (Chagas' disease) acquired in South America. A first-rate account of the Darwin-Chagas' hypothesis has been given by Jared Haft Goldstein in his article "Darwin, Chagas', Mind and Body" (*Perspectives in Biology and Medicine* 32 (1989): 586–601).

By 1910 there was a growing realization of the role of blood-sucking insects in the transmission of microbial and parasitic infections. In 1896, Ronald Ross showed how the malaria parasite was transmitted by mosquitoes. That same year, in Africa, David Bruce began to implicate the tsetse fly in the transmission of trypanosomes. In 1901, Major Walter Reed of the U.S. Army and his colleagues demonstrated the transmission of the yellow fever virus by the *Aedes aegypti* mosquito. It had become a standard experimental practice of the field-working microbe hunters of that time to dissect any blood-feeding insect new to them to see what in the way of microorganisms might be lurking inside.[4] In this way Chagas dissected some of the reduviid bugs of Minas Geraes. Under the microscope, he saw protozoan flagellates in the bugs' intestinal tracts that looked like an intermediate, insect-dwelling stage of a trypanosome.

In his bush laboratory, Chagas had come to the end of his technical resources. He did not have the laboratory-bred animals to attempt the experimental infections necessary to determine whether the bugs' flagellates were an intermediate stage of a parasite of mammals or an organism confined to bugs alone. Chagas sent some of the bugs to Cruz in Rio. Cruz fed them on a marmoset. Three weeks later large numbers of strange-looking trypanosomes, morphologically unlike the familiar trypanosomes of African sleeping sickness, were present in the monkey's blood (figures 1A and 1B).

More experimental animals were offered to the bugs, and guinea pigs, rabbits, and puppies became infected. It was

4. Reduviid bugs are still dissected but for a different, diagnostic purpose. In late chronic Chagas' disease, the motile(trypanosome)-stage parasites become very scanty in the blood and are difficult to detect either directly or by test-tube culture. Reduviids are bred in the laboratory and these clean bugs are allowed to feed on the patient. The *Trypanosoma cruzi* trypanosomes, even if present in very scanty numbers undetectable by microscopical examination, will colonize the insects' intestines and be readily identified. This diagnostic method, called xenodiagnosis, is very sensitive but it does not give rapid results; it takes about a month before the bug is "ready" by which time the patient too often has fled or died.

FIGURE 1A
African Trypanosome

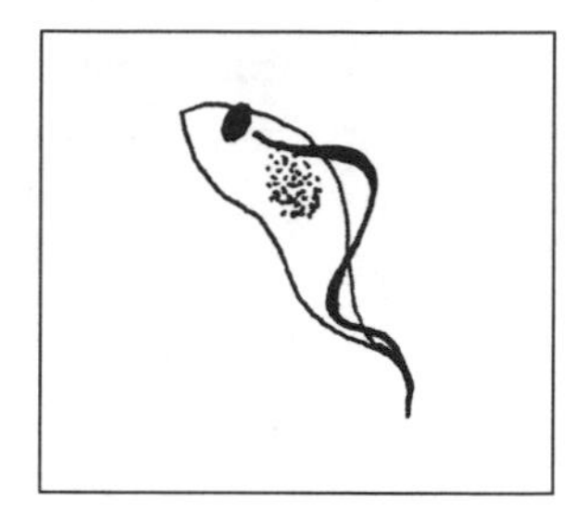

FIGURE 1B
Marmoset's trypanosome

obviously a parasite of mammals and was broadly specific—it could infect many different species. What's more, the parasite was pathogenic—some of the infected animals sickened and died. But did it infect humans?

When the news of the experimental infections reached Chagas in Minas Geraes, he returned to the village where he took blood samples from resident cats and dogs. In the blood of a cat Chagas found the trypanosome, an organism that he was later to name *Schizotrypanum cruzi* (later renamed *Trypanosoma cruzi*) after his admired Master. A few days later Chagas found the trypanosome in the blood of a sick three-year-old girl (who turned out to be one of the lucky survivors; she was alive and well 60 years later). Blood samples from other children were examined in an attempt to connect the trypanosome with a specific disease. Some but not all of the children with the trypanosome in their blood were anemic and stunted in growth. A few children with many trypanosomes in the blood gave a history of convulsions, and these children, Chagas found, died of a neurological disease within weeks or a few short months after he first examined them.

Two years later, in 1912, Emile Brumpt, the parasitologist of the Paris Pasteur Institute, carried out research in Brazil that proved transmission was *not* by the injection of the trypanosome from the bite of the bug (like all other known arthropod-borne parasites) but by a contaminative route from the bug's feces. One sleeps. The reduviid bug steals from its hiding place and sticks its sticker and sucks up blood. As it sucks, it defecates an excrement containing the infective stage of the trypanosome. Through either a skin wound made by the bug or an existing abrasion, the parasite enters. The sleeper scratches at the offending bug and the trypanosome enters. The bug commonly feeds near the eye. The excreta is rubbed into the conjunctival membranes and soon there is a swelling of the eyelid.

In the next few years two more crucial facts of the American trypanosome's biology came to light. Most extraordinary, the trypanosome entered a variety of tissue cells where, as an intracellular parasite it rounded up to become a leishmania form. The leishmania forms divided asexually to form nests in the tissues. It was like a leishmania but not like a leishmania because it wasn't restricted to infecting only macrophages. Many cell types could be invaded although the predilection was for heart and nervous tissue cells. So here was, much to the surprise of the parasitologists, a unique Hemoflagellate of the Third Way with *both* trypanosome and leishmania stages existing simultaneously in the infected host. Nothing like had been seen before—or has been since. It was purely American. Its family tree is still a mystery; we still do not know from where, when, or protozoologically from whom it evolved. Using the very sophisticated modern techniques of DNA homologies, Herman Heckler of the Swiss Tropical Institute in Basel (even the Swiss have a tropical institute; of the western world only the Americans don't have such a research center) addressed the problem and came up with a paper entitled "Man and Sea Urchin—More Closely Related

Than African and American Trypanosomes" (*Parasitology Today* 9 (1993): 57).

The second fact was that the laboratory animal experiments and Chagas' Minas Geraes cat had predicted the epidemiology of *Trypanosoma cruzi* in the real world. Chagas' disease was proven to be a zoonosis. Many wild and domestic animals were found to be naturally infected, the armadillo and opossum being particularly important reservoirs. I'll tell you about the Chagasic racoons of Maryland when we reach the chapter on our century. However, since we are still in the pre-Columbian period, let us return from this rather long diversion to the Inca in the basket.

She had lived in Cuzco, Peruvian capital of the Incas. Death came when she was about 20 years old, and keeping to the custom of the time, her corpse was placed in a long wicker-like woven basket. Almost 600 years later a team of scientists led by Gino Fornaciari of the University of Pisa's Paleopathology Laboratory and Elsa Segura of Argentina's Fatala Chaben Institute made the long-delayed diagnosis of Chagas' disease as the cause of death. The mummy's viscera still showed the late stage Chagas' mega syndrome. The heart and intestinal tract were greatly enlarged. The dilated colon held an enormous amount of fecal matter. Amazingly, the parasites were also present, albeit in a mummified state, and could be identified by electron microscopy and the unerringly specific staining developed by monoclonal antibody reagents for *Trypanosoma cruzi.* Chagas' disease in Chilean mummies had been suggested, but from this unfortunate lady of Cuzco, Fornaciari and his colleagues noted that this was "the first direct demonstration of this disease [Chagas' disease], and the agent causing it in South America during the Inca empire immediately before the Spanish conquest of the continent."

The story line of this chapter has been rather like a Nintendo game, "Get the Indian to Argentina." You would have to jockey the brave bands of Amerinds down the continents to Tierra del Fuego all the while beset by microbial monsters. Some tribal groups in the game might go blooie, but most would get through to safe havens on the way. Those pre-Columbian Amerindians had their medical troubles, but by and large they were not too bad off. The Americas, in fact, were relatively healthy until that disastrous morning of October 12, 1492, when three sailing ships appeared on the horizon of the Bahamas.

Chapter 3

Who Gave Pinta to the *Santa Maria?* 1492 to 1635 A.D.

THE *SANTA MARIA* was hardly the *Love Boat*, and after a month at sea Juan de Morguer had a acute longing in the loins. He had had a sweetly satisfactory night before departure with his wife; for the sailor whose every voyage was so perilous, sex before sailing was like the last rites. That next day, August 3, 1492, the caravels *Nina* and *Pinta* and the ungainly cargo vessel *Santa Maria* (on which Juan was an able-bodied sailor), which carried them on this mad enterprise, put out to sea from Palos. The wind was fair, the voyage fast, and by August 9 they reached their first destination, the Canary Islands. It was the intention of that jumped-up Italian (some whispered that he was Jewish, despite his almost fanatical Catholicism) who styled himself the Admiral of the Ocean Seas to spend only a few days in the Canaries taking on water and provisions before putting forth to the unknown void of the western Atlantic, an unexplored route that the Italian had convinced the blessed Queen Isabella would take them to Japan and Cathay, the fabulously rich dominion of the Great Kahn.

The *Santa Maria* was such a sow of a boat to sail, slow and constantly breaking down. Well, what can you expect from a rental? A cranky rudder delayed them for a month in Gomera (an island of the Canaries). Finally departing Gomera on

September 6, they were becalmed offshore for three days until a westerly wind blew. For the fleet's 90 crew members the delay was a welcome respite.

Day on day the ships sailed west. The brown-skinned harlots of Gomera had been of great comfort, but now almost two months out of the Canaries the crew was becoming as cranky as that damned rudder, close to mutiny and forcing the admiral to come about and return to Spain; they would all die on this endless ocean. A few days more, Columbus cajoled; the first to sight land would be given a prize of gold. Today, in fact, there were signs that the fabled Indies were near. Land-based sea birds flew overhead; driftwood and green plant debris floated past the ship.

Sailors' legend had it that the women of Cathay were beautiful and willing beyond compare. Juan de Morguer was not like that murderer Barotlomé de Torres whom the queen released from prison when he volunteered to sail on this suicide voyage. Bartolomé had practiced odious sex in prison and on this voyage he had sodomized one of the ship's boys, the grimete Diego Bermudez. This was not Juan's way although sodomy was common among sailors of all countries. Besides it was dangerous in this time of the Inquisition when sodomites were burned at the stake. There were no priests on this voyage, but Columbus was more catholic than the Pope. An English sailor he once met told him that the British navy had hanged more sailors for sodomy than for murder and mutiny.

Such were the sexual musings of Juan de Morguer on October 12, 1492, when the lookout, perched high on the main mast, shouted "Land." The *Santa Maria,* together with the *Nina* and *Pinta,* now made her way to the still misty, distant shore. They came finally to a tropical beach. On the shore to greet them were the Cathays, the Chinese, subjects of the Great Khan, brown skinned, as they were supposed to be, but hardly adorned with rich fabrics and jeweled ornaments. Mother of God! These people were stark naked! The admiral may have been disappointed by the absence of material wealth but Juan de Morguer was exalted. He had died and gone to heaven; Paradise was on the other side of this vast, trackless ocean.

They had come to an island (probably San Salvador, the Bahamas) whose inhabitants spoke no language known to the ship's interpreter, Luis de Torres, who was once a Jew and first tried to communicate with these Indians in Hebrew. Eventually, de Torres came to learn that these people called themselves Arawaks and that they knew nothing of the Great Khan, gold, or Jesus Christ. For Juan and his shipmates these heathens of the Indies were childlike, their women delightfully accommodating, the reluctant few soon persuaded by a little force.

For more than three months the Spaniards explored these waters so rich in islands and so poor in gold and treasure. Some islands were inhabited with people not so simple or naked as the Arawaks but neither were they the Indians of the Khan. Their chiefs spoke of great kingdoms to the west but it was all mythic rumor. Finally, on January 16, 1493, the admiral turned his ships to the east and began the voyage home to Spain.

It was a good voyage, free of the outward bound apprehensions. Juan was not made wealthy by the trip, nor was the admiral, nor would be Queen Isabella. At least Juan had the torrid carnal memories of those many brown-skinned women. During the second week at sea these sexual reveries were interrupted when he noticed a small, round, somewhat hard lesion on the head of his penis. It caused no pain and he didn't worry overmuch; his penis had been through the

"Venus wars" of the sailor—a wife at home and a strumpet in every port. By the time they put into the Azores (it was Portuguese and they were a bit off course to their home port of Palos in Spain), the lesion had healed and Juan was no worse for wear, so to speak.

Still off course, the admiral made an embarrassed landfall at Lisbon, and only after some considerable argy-bargy did the Portuguese allow him to sail south to Spain. The time and dalliances in Lisbon were pleasant enough but Juan felt great joy when they put into Palos on March 15, 1493. He and his mates were heroes, men who had come back from the edge of the world. Juan would voyage no more; God and the blessed Virgin had brought him back unharmed to his family. He would live out his life in the sun of Palos, regaling all who would listen to his story of how he and the admiral sailed west to the Indies and discovered a new world.

At the age of four my son had a flash of theological insight and declared "God tricks you!" God tricked Juan of the life in the sun of Palos. His serpent from the Eden of the Indies, coiled and invisible to the eye, was even now a predator in the brain and other body tissues of Juan de Morguer.

About three months after his return Juan had a fever. He felt unwell and his skin had a weepy rash. Everyone in Palos, especially the seafarers, had their share of fevers and fluxes. Sailors, long at sea, came down with a condition they called scurvy that gave them a rash and bleeding gums so tender that it was an ordeal to eat their hard, rotten food. The Palos doctor said this wasn't scurvy; it looked to him more like ringworm. Olive oil was applied, then vinegar, then secret unguents, all to no avail. Juan began to have piercing headaches; strange thoughts plagued his mind, and his vision began to blur.

Two years after his return this once hero of Palos, the

man who had been in the company of the admiral, now was a pitiful object of derision. Taunting children followed him, mocking his strange gait; they called him El Gato, the cat, as he threw his legs out like a drunken tabby in his walk. He uttered strange noises and screamed at the demons and monsters tormenting him. Lightening pains would strike his legs and he would fall, helpless, to the ground. On the last day of September 1495, the heart of Juan de Morguer burst (an aortic aneurysm) and the man who sailed with Columbus died—mad, blind, and syphilitic.

Lusty fifteenth-century Europe was a tinderbox that syphilis ignited. With stunning speed a killing epidemic of syphilis descended a year after Columbus returned from the Americas. It is conjectured that the spirochete mutated, when it reached Europe, into a highly virulent, invasive form. Fifteenth-century health statistics are not accurate, but even then the tidy military maintained reasonably good accounts and from them we can gain some insight into the devastations of that first epidemic of the "French pox." The epicenter was Italy, but it was against the assaulting French that syphilis joined the order of battle.

In 1494 Charles VIII of France thought he should own Naples by reason of some convoluted inheritance from his father who had married a claimant. So he marched through Italy and besieged Naples in February 1495. There his conquering troops disported themselves and acquired the "new" syphilis. They became so debilitated by the disease that only feeble resistance could be mounted against the league formed by Ferdinand V of Spain, Pope Alexander the VI, and Maximillian I of Bavaria, along with some help from Venice and Milan. Charles's army was not pure French but rather a polyglot assemblage of troops contributed from his princely friends of other states and mercenaries. After losing the

Naples war, this multinational force dispersed, and its members seeded all of Europe with syphilis. Within ten years England, Germany, Russia, Poland, and the Scandinavian countries had become, in the epidemiological sense, syphilitic. Vasco da Gama is believed to have introduced it into India during his 1498 around-the-world cruise.

Low born and high were equally affected. Charles VIII, that "first contact," had it as did the clergy up to and including the Cardinal Bishop of Segovia and His Holiness, Pope Alexander Borgia. Syphilis made Ivan the Terrible even more terrible. This first Muscovite tsar (otherwise quite a nice person), crazed by the spirochete, whacked his son with an iron bar. Another son, Feodor, believed to be a congenital syphilitic had the chutzpah to propose marriage to Queen Elizabeth I but she turned him down cold; not only was he quite loony, he was already married at the time of his proposal. Thus it may be, as several historians have suggested, that syphilis made a deeper impression on the history of Renaissance Europe than any other disease.

There is a nice touch of drama to our fictionalized account of Juan de Morguer and his American souvenir of syphilis. Juan was a real person but is our story true to medical history? The syphilis transactions have long been hotly debated among those interested in the historical epidemiology of infectious diseases. Did syphilis come from the Americas as I and others believe, or was it present in Europe, probably originating in Africa, for many centuries before 1493?

In entering the continuing syphilis debate of who gave what to whom, we must first consider the disease and its causative organism. To be confusingly precise, syphilis should be syphilises since what seems to be a single organism with four different names causes four different kinds of disease.

Under the microscope or by the usual immunological tests of identity, the spirochetes cannot be distinguished one from another.

First there is yaws, caused by a spirochete given the name *Treponema pertenue.* It is nonvenereal, transmitted by contact other than sex, and it can be quite horrible. The infection begins as an "unimportant" papule which later breaks down to become great destructive lesions of the skin and underlying bone. It is thought to have had its origins in hot, humid sub-Saharan Africa and spread to many parts of the tropics, including the tropical Americas, with the slave trade. There seems to be a cross-protection between yaws and true syphilis—if you can call bone destruction on one side and dementia and death on the other side "protection." The contemporary good news about yaws is that if anything can be called a miracle cure, it is penicillin's action on *Treponema pertenue.* Within days of a single injection of penicillin, the yaws lesions begin to resolve. When I began research in Papua New Guinea in 1962, there were numerous cases of yaws in the villages of the Sepik plain. In 1963 and 1964 there was a massive penicillin treatment campaign in the region, and when I returned in 1964, not a single case remained; not a single lesion in children or adults could be seen. Remarkably, in 1990 the Sepik was still yaws-free.

In pre-Columbian times when the Indian men of Hispaniola were overcome with unrequited sexual desire, they would try to catch amorphous spirits who, when captured, would be transformed into beautiful women. The most successful catchers of these slippery, elusive spirits were men with horny hands. It's a myth, but for medical historians even myths are the stuff that clues are made of. These historians have interpreted the horny-handed Indians as having Pinta, the second of our spirochete diseases, and it as having had its origins in the tropical New World. The Pinta spirochete is named *Treponema carateum,* but again it is a name of convenience since all four spirochetes are so alike as to be considered variants rather than authentic species. Pinta has remained an American disease, present in the region between Cuba and Mexico, through Central America to the Amazon Basin. Pinta is confined to the skin where it causes open sores. Transmission from person to person is most probably by direct contact with these lesions. In the later stages the skin becomes depigmented and/or with blue-black splotches, and keratinized, maybe the cause of the horny grasp of the mythic Indian.

The third spirochete-treponeme is also a member of the syphilis family, although also a somewhat unorthodox one. The disease it causes goes by several names—endemic syphilis, nonvenereal treponematosis, and its Arab familiar, *Bejel.* The microbe itself has been named *Treponema pallidum* II, as if it were a son named after a distinguished father. *Bejel* is considered to be a nonvenereal form of syphilis that is spread, like yaws and Pinta, by direct contact or possibly through shared use of eating utensils. It occurs mainly in children. It is of Old World origins, and although once present in Europe, its present landscape epidemiology is the sub-Saharan region of West Africa. It is not a nice disease, beginning with lesions of the mouth and skin and progressing to great destruction of the skin, crippling destruction of the long bones, and grotesque disfigurement as the nose and mouth are destroyed. Unlike true venereal syphilis, the central nervous system does not become involved in *Bejel,* nor is there congenital infection in which the spirochete of true syphilis is passed from a mother to her fetus.

True syphilis is *Treponema pallidum.* The virulent course of venereal syphilis was enacted in the body of Juan de Morguer. He experienced the progression from the primary to tertiary stages: from a small skin lesion, to skin rash with weeping papules, to fatal cardiovascular and neurological involvement. Juan de Morguer's novelized case history is textbook syphilis but the true character of true syphilis is its variability. It may not fulminate as it did in Juan de Morguer's body; instead it may lie dormant, clinically quiescent for many years or even during the entire life of the infected person. It is still not known why *Treponema pallidum* is so capricious in expressing its virulence. Strain differences? Spontaneous mutations? Host differences in the genetic-immune constitution between individuals or groups? And why was it so virulent, so lethal, in the great European syphilis epidemic of 1493 to 1510? Was it benign in pre-Columbian tropical America and became syphilis as we know it today only when it emigrated to the Old World? One clue may have been provided by a hunter-gatherer group of Indians living in pristine isolation within the depths of the Amazon rain forest.

In 1970 a group of American epidemiologists led by Richard V. Lee of Yale University went to the Amazon rain forest in search of the syphilis family of diseases in the Indian tribes that had had little contact with the outside world. They did what infectious disease epidemiologists do when they are in the field far from their laboratory resources: they drew blood and preserved the serum for later serological testing, they examined the people for signs of disease, they asked questions. It must have been a disappointing exercise because in no instance did they find any overt evidence of venereal syphilis or any other treponemal disease in these Indians. They expected the serological findings to confirm this freedom from disease and I can imagine their surprise when the lab reported the unexpected and exciting results of high positivity rates for antibody to the treponeme in the Kayapo tribe, rising with age to 90 percent in the 40+-year-old group. In the absence of treponemal disease or even history of this disease, Lee and his colleagues conjectured that they were seeing the serological "footprints" of a new, previously unrecorded, avirulent treponeme of humans. The other possibility offered was that this unseen (they could not isolate the organism due to technical limitations) microorganism was the last, isolated remnant of the original *Treponema pallidum,* the patriarch syphilis spirochete. Here in this Indian tribe, sheltered and isolated within the near-impenetrable jungle, this American treponeme would have been passed, unchanged and benignly adapted to its human host, from one generation to another. But if this were true where did *that* spirochete-treponeme come from?

I called Dick Lee, now Prof. Richard V. Lee at the State University of New York at Buffalo, School of Medicine. We traded "war" stories of our old field researches, and I guess he was surprised to learn that I was a graduate of the University of Buffalo before it became a state institution. I asked him whether he had any new thoughts or even old new thoughts on the mystery spirochete of the 1970 Kayapos. He said that he still believed that it was a benign spirochete and the ancestor of venereal syphilis microorganism and therefore that syphilis was of American origin. But, he said, there had been no new work; no one had gone back to follow up on that original study and, regrettably, with all the research money in the hands of the molecular mavens, there were no funds for the kind of richly rewarding field studies of the 1960s and 1970s. And he still wondered where the Kayapo spirochete had its origins.

Astrophysicists have pretty much solved their origins problem. They can trace back those many billions of years to coalesce the contents of limitless (?) space into the unified mass that had big-banged into the creation of the universe. Microbiologists have a more difficult challenge in searching for the prime origins of the organisms that populate the microscopic or submicroscopic world. However, through powerful DNA technology ancestral relationships of microorganisms from past and present are beginning to be illuminated. Even the treponemes may yield their ancient secrets to DNA probes. True they can't be cultured en masse. They still have to be grown experimentally in the testicles of rabbits, to allow extracting a dollop of their DNA; but through the new amplifying technique called PCR, the DNA from just a few organisms could be used to characterize their homologies and affinities. That's not happening yet. Researchers have to go, like bank robbers, to where the money is, and there isn't much cash deposited in the syphilis family's account.

Even those theoreticians, the armchair microbiologist-epidemiologist speculators, cannot offer a plausible evolutionary path for *Treponema pallidum.* And we certainly can't fathom how it became a parasite of humans in the New World before infecting the Old World. If it were a zoonotic microbe of New World monkeys, we could at least put it on the evolutionary path.[5] But a treponeme of any sort has never been found in a New World monkey of any sort. It may be that there is or was a treponeme in some neotropical animal, other than a primate, that once made that great biological leap to humans. It can happen. For example, DNA homologies have recently revealed that *Plasmodium falciparum,* the

5. New World monkeys (Plattyrhines) are very different from Old World monkeys (Catarrhines). For one thing, Plattyrhines can swing by their prehensile tail and Catarrhines can't. That "African" monkey in the movie *Outbreak* was actually a South American. Forgetful Hollywood.

most lethal malaria parasite of humans, did not come up from the ape (as did other species of human malaria parasites) but rather jumped from birds to humans. And once in humans it became fastidious for our species; it can no longer infect birds, and only a few species of primates, such as the owl monkey, can become experimentally infected. Is there, was there, an armadillo, macaw, or capybara with ancestral syphilis somewhere in South America?

Chapter 4

Ten Little Indians and Then There Was One: 1492 to 1635 A.D.

WITH THE POX and plagues in Europe, and with the conquistadors' smallpox and adult-killing diseases of childhood in the Americas, the peri-Columbian fifteenth to seventeenth centuries were a perilous time for the survival of the human species. The epidemiological ground for killing epidemics had been fertilized several centuries earlier by an almost global change in human behavior and social organization. I am told that every individual plant of every species of bamboo, no matter where they are located, will go to seed and then die at the same time. One bamboo can be in Manila, another behind my old lanai in Honolulu; driven by some genetic calendar, both will flower, seed, and die within days of each other. Like the bamboo, the peoples of Europe, Asia, and the Americas, noncommunicating but as if driven by a common time-controlled, species-inherent force, began to change from the small-band nomadic life to become sedentary; to assemble in larger groups within villages, then towns, and then cities; to develop agriculture and domesticate animals. There was more food and more fecundity; human numbers began their progressive ballooning in a period we have come to call the early Middle Ages.

Many components make up the epidemiological package of an infectious disease; the size of the population under attack is one important factor. For example, the normal human child reacts with immunological vigor to the measles virus and that pathogen's persistence is relatively short. It therefore needs a good supply of new recruits to keep it circulating permanently in a community. A population size of about 200,000 is needed to provide a constant inventory of new nonimmunes and keep the measles virus going. Other microbes have come, for reasons not well understood, to a different arrangement with their human hosts. They neither kill (or at least kill swiftly) nor are killed by the immune response. These are indolent microorganisms that persist for long periods; they can wait for new susceptible people to infect, and because they can, only small population groups are needed to perpetuate them. The herpes simplex virus and the tuberculosis mycobacterium are examples of small-group infections. Finally, there are microbes independent of human population size for their long-term survival. These are the zoonotic microbes whose constituency is the animal population. Humans get the infections from these animals, and they are not necessarily wild animals; many domestic animals serve as zoonotic reservoirs. Rabies, yellow fever, and African trypanosomiasis (sleeping sickness) are a few examples of the many zoonoses.

These epidemiological ingredients had been percolating in Europe when the plague struck in 1348 A.D. It was not a good year to be in Europe; crops had failed and the rural, hungry poor came flooding into the towns and cities unprepared to house the immigrants, ignorant in ways to meet the added sanitation demands, and with quite primitive sanitation facilities to begin with. The brown rat, *Rattus rattus,* also a city dweller, literally fed on these conditions so parlous for humans; it thrived and multiplied.

It is believed that the plague bacillus, *Yersinia pestis*, came to Italy from Asia in a ship's rat. The more specific opinion is that it was brought to Europe from Constantinople by the returning crusaders. By crusaders' vessels or by merchants' ships, the Italian brown rat rapidly contracted the infection, transmitted from rat to rat by the rat flea *Xenopsylla cheopis*. If anything, the brown rat is even more susceptible to *Yersinia pestis* than the human is. Dr. Stephen R. Ell, a pestilence scholar, noted that "Humans are incidental, if spectacular victims of plague." And so, when the plague struck, the rats died like flies. The rat fleas, denied their regular source of food jumped to the "big rat"—the human. Now the humans began dying like flies. When winter came, there was an ominous change in transmission from the flea route. Huddled together for warmth and safety in their airless hovels, the humans inhaled the bacillus suspended in the air. There was now direct transmission from human to human without agency of rat or flea. The germ entered the lungs and became even more lethal; the bubonic plague had become the pneumonic plague, which was terrible.

After several years it was over; the Black Death was a burnt out epidemic, burnt out but not completely extinguished. After smoldering for almost three centuries, it returned with familiar savagery. In mid-seventeenth century, Europe was again with plague. It came to London in 1665 with such ferocity that the city was nearly abandoned; the anguish of those plague years was vividly described by that master storyteller, the creator of Robinson Crusoe and Moll Flanders, Daniel Defoe. Then, helped by the strict imposition of quarantine laws, that plague too fizzled out. After the seventeenth-century outbreak, the plague left Europe, seemingly forever. Nevertheless it left its long-residual demographic mark. It has been calculated that the two epidemics so reduced Italy's population that it took almost 400 years before Italy's population had recovered to its preplague, thirteenth-century size.

Ell and other historian-epidemiologists have proposed a set of intriguing explanations to account for the whereabouts of *Yersinia pestis* during those smoldering-slumbering 300 years. Explanation 1 claims the brown rat that died was replaced by the more immune-tolerant Norway rat *(Rattus norvegicus)* that lived and so kept the cycle pretty much confined to the rodent population. Explanation 2 has the plague bacillus going to the domestic animals—dogs, cats, pigs (Boccaccio has a tale of pigs dying after coming in contact with plague victims) which would act as "blotters." Of course, it is reasonable to suppose that those domestic animals would act as a threatening reservoir to humans, an infectious menace of propinquity. Explanation 3 (or 2½) has *Yersinia pestis* going into wild animals and its maintenance in those animals as a sylvatic cycle until human-induced ecological perturbation brought conditions of renewed contact.

Nor did all infected die of the plague. There seems to have been a pattern of mortality and morbidity. More men than women died. More adults than children died. Although rich and poor were infected equally, the rich died in greater numbers than the poor—a fact that made the perpetuation of nobility a dicey proposition during the plague years. Ell believes a common cause underlies these disparate epidemiological statistics—iron. Most pathogenic bacteria love iron: they thrive on it; they go reproductively wild on the iron diet. They are the microbial Popeyes. Iron-deficient blood can act like an antibiotic and, paradoxically, healthy "full-blooded" people may well have more severe microbial infections than the anemic. Several misfortunes of good charity have attested to this; when malnourished, anemic peoples, victims of wars, famines, or other disaster, have been given, by aid organizations, a high-caloric, iron-rich diet, there has frequently been an exacerbation of illnesses such as tuberculosis, dysentery, and even malaria.

Iron in the blood and the lack of it can be a logical explanation of the plague statistics. In medieval Europe women were secondary to men in the "food chain"; they ate last and less. Moreover, premenopausal women of childbearing age are more likely to have lower blood iron levels than men. This would make men more susceptible than women and give them a higher mortality rate. The rich eat of a richer fare than the poor—lots of meat and vegetables for the wealthy, a bit of bread and gruel for the peasantry; iron-red blood for the rich, anemia for the poor; death from plague as the iron-fed bacillus multiplies in the blood of the rich, survival of the poor as the bacillus starves from lack of essential nutriment.

Frankly, this is more of the plague than I intended but it's easy to get carried away by the plague and it does provide a useful insight into the intricacies of infectious disease epidemiology. Also, after 1492 every disease, every disaster, every political upheaval in Europe affected and afflicted the New World.

There is plague in America today, a constant sylvatic zoonosis that occasionally reminds us of its presence by infecting a hunter or other lover or resident of the wilderness. During a recent trip to the Centers for Communicable Diseases Laboratory in Fort Collins, Colorado, a former graduate student from the University of Hawaii, Dr. May Chu, took me on a tourist tour of the area. May is a microbiologist with research interests in zoonotic diseases, so tourist touring with May is more than mountain vistas and bugling elks. On the rather barren prairie below the mountains, she pointed out a continuous series of mounds that are prairie dog towns. "You see that colony's town," said May. "That's a dead colony; all the prairie dogs died of plague. They are as susceptible as rats and the plague usually wipes out the whole colony. Now that colony over there is still alive—no plague yet." We passed dead prairie dog town after town. I was astounded—so much plague. Could the sylvatic plague spill back into the human population? "Maybe, if there was a kind of apocalypse, a breakdown in our social structure," mused May. Then we drove by a new housing development hard by the town of the now semidomesticated prairie dogs and there was a further bit of May musing. "Well, it could also come back into humans from that (finger pointing to the new houses on the prairie) sort of change bringing human and reservoir animal together." Finally, on the way back we speculated on the origins of the plague in America. In earlier times did the plague bacillus come from Europe and did it play a role in the post-Columbian collapse of the native American population?

Population size and structure are all important in epidemiology. The people number must, as a demographer friend's T-shirt message once proclaimed, be broken down by sex and age. However, there was no census of the Amerindian population on October 12, 1492. Population estimates differ widely and wildly, from 8 to 100 million people (estimates cluster at 40 to 70 million) in all the Western Hemisphere with 250 thousand to 5 million in North America. We do know that there were "ten little indians" before the conquistadors came and only one or two 50 years afterward.

As in medieval Europe, a centripetal movement of population coalesced into large groups. Small bands of nomads, swidden agriculturists, and hunter-gatherers persisted, but there were also aggregations of peoples in villages, towns, and cities. As a perspective: In the thirteenth century, Paris with a population of 100,000 was the largest city in Europe; but in the same era, Cahokia, a city in Illinois, had 40,000 people. As in Europe with population growth and coalition, in the Americas there came new risks from infectious diseases. Also we know with sad certainty, there was a holocaust-like collapse of these American civilizations within one generation of the coming of the Caucasian. If it were any animal other than the human, we would nowadays put the sixteenth-century Amerindians on the endangered species list.

Early-colonial Ecuador is a good example. The English geographer-historian Linda A. Newson has noted that the coastal peoples were most affected since they served as first contact with the Spanish. She calculated that in Guayaquil the rate of decline was 60 to 1; if Guayaquil had 6,000 people in 1520, only 100 would have been left in 1570. In Mexico, a hypothetical village of 250 people in 1532 would be reduced to 10 people in 1600.

When we seek the cause or causes for the post-Columbian near extinction of the Amerindians, the "clean" clinical answer comes first to mind; it was an infectious disease introduced into a "virgin soil" population. Undoubtedly, new infections in the immunologically naive can bring a high mortality, but virulence alone cannot fully explain such a swift and massive depopulation. I can think of few microbial pathogens that would kill so greatly and so consistently for so long a time. Even the notorious Ebola virus epidemic in Zaire ended spontaneously after killing "only" 244 people of the "only" 315 people known to have been infected.

We don't have to look to an exotic to account for a massive human die-off. The greatest global epidemic (pandemic) ever experienced, of which one commentator said "no war, no famine has ever killed so many in so short a time," was the Spanish influenza outbreak of 1918. It killed 20 million people, 500 thousand in the United States. In hard-hit Philadelphia so many died that coffins were in short supply and corpses had to be buried in blankets. It was a very virulent strain of flu virus that could be fatal within three days of the onset of symptoms. A flu like that could have wiped out the Amerindians. However, Europeans and natives should have been affected equally, and there is no record of an abnormal number of deaths among the conquistadors. The Europeans might, somehow, have been more resistant. Possible but not probable.

Nor do we know the effect of the diseases of childhood on a virgin soil population in 1550. Measles, pertussis, and chicken pox cause great havoc in both children and adults when first experienced by isolated populations—Eskimos, Pacific islanders, hunter-gatherers in remote rain forests. However, even for these peoples the new infections weren't Armageddon. A pertinent look back is the work of James Neel and his colleagues who studied an outbreak of measles in the Yanomama indians, a tribe living deep in the jungles of the Orinoco River basin of Brazil and Venezuela. When Neel's team visited them in 1970, they had had little or no contact with people from outside their territory and therefore no experience of the diseases of civilization—measles, mumps, tuberculosis, and so on. Then in 1968 a missionary came to them. The missionary brought his son; the son had measles. The infection spread like wildfire through the Yanomama, striking young and old alike. Those who could fled back into the jungle, but most were too sick to move from the settlement where the Yanomama met the missionary. There were many deaths when compared to the mortality rate from measles in a "normal" population. About 25 percent of the infected Yanomama died. A measles vaccine was tried to stem the outbreak, but that 1968 preparation did more harm than good. Eventually, the epidemic burned itself out without seriously threatening the existence of the tribe.

Yellow fever could have done it. Africans—yellow fever carriers—were brought to the New World as slaves early in the colonial game. Even that renegade, Hernán Cortés was accompanied by black slaves when he assaulted the Aztecs and took their king, Montezuma, hostage in 1519. But yellow fever doesn't quite kill enough to be the guilty party, nor would it be selective in just killing the natives and not the Spaniards.

Malaria, which was also introduced early in the colonial period, doesn't kill with a high enough percentage either, but we shall come back to malaria shortly.

Smallpox is a pretty good candidate. It spread into the New World from Jesuits and their need for converts. Jesuits who came from Spain to proselytize the heathen Amerindians had among them those who either had active cases of smallpox or were carriers of the smallpox virus. The Jesuits dragooned large numbers of natives to work and be converted. With this critical mass, epidemic smallpox was common and many Indians died (but now, as Christians, they went to heaven). The Jesuits, acknowledged masters of deductive reasoning, were aware of cause and effect; they arrived and the natives died. They came to know that they were the carriers of disease and sometime during the early seventeenth century they were instructed to carry a hoe with them when they came to the Americas, not to plant seed but to "plant" the dead they left in their missionary wake.

Smallpox, a hardy virus, can survive for quite a long time in fomites. Certainly it was a major wipeout pathogen after introduction into the Caribbean in or about 1516. The disease is believed to have assisted Cortés in his conquest of Mexico by debilitating the Aztec warriors. Curiously, it is said that the epidemic of smallpox only affected the Aztecs, leaving the Spaniards unaffected. Given the close contact fighting (the Aztecs wanted to take the Spaniards alive for ritual sacrifice, whereas the Spaniards simply wanted to kill Aztecs, so the story goes) and the highly contagious nature of smallpox, the virus would not have spared either combatant unless a good many of the Spanish troops had had the disease and recovered from it in their youth. At any rate, smallpox, as bad and as ugly as it is, isn't sufficiently lethal to reduce a population from 50 to 1 in fifty years.

Plague might have been the culprit. The time when Amerindian populations were crashing just preceded or was coincidental with the great seventeenth-century plague epidemic of Europe. However, plague would, again, kill both parties equally. As a matter of fact, the colonial governors as well as Cortés in their reports to the king and his administrators commented on the robust health of their troops in the New World.

So the mortality numbers for any one infectious disease do not add up to the 60 to 1 or even the 25 to 1 fall. The cumulative effect of several coincidental virulent infectious diseases would come closer to the extinction statistic. But personally, I believe that the *coup de grace* was the statistically intangible factors of despair and demoralization. The Amerindians were, quite suddenly, beset by strange and fearful illnesses at the same time as they were being brutalized and enslaved by the Europeans. Their civilization was collapsing, their familiar gods replaced by a strange white people's god nailed to a cross. A people do not prosper or readily breed nor do their children survive under such oppressive conditions. And they panic.

We have no good descriptive anthropological records of the sixteenth- to seventeenth-century Amerindian dissolution, but panic is panic in any century, in any people faced with overwhelming disease and oppression for which they do not know the cause or cure. Thus, an 1877 measles epidemic in the virgin soil population of Fiji is of a piece with the plight of the sixteenth-century Amerindian. Here is how an English physician, W. Squire, who was at the epicenter of the Fijian measles epidemic described the collapse of village life: “Excessive mortality resulted from terror at the mysterious seizure, and the want of the commonest aids during illness; there were none to offer drink during the fever, nor food on its subsidence. Thousands were carried off for want of nourishment and care as well as by dysentery and congestion of the lungs. We need invoke no special susceptibility of race or peculiarity of constitution to explain the great mortality.”

Ancient history, remote in time and place. It can’t happen here. Besides, natives were always vulnerable to the forces of civilization. And yet, when we look at the present, looking backward is cold comfort. Today’s plagues are more deadly than those of the old centuries. There is Lassa, Ebola, and Marburg of the swift death; there is the HIV of AIDS that brings an inexorable, inevitable death. These are the notorious diseases of our time. These are the emerging pathogens that seem to have come from nowhere and whose original sources remain unknown or poorly understood but are believed to have crossed the species barrier to humans after the habitat of the natural animal hosts had been degraded. These are really bad-news viruses. I don’t think, however, that they portend the extinction of the human species. One way or another they would be contained, by harsh measures of segregation when there was no cure or prophylaxis. These are not the viruses that would bring us to the brink of extinction as the "new" pathogens of the sixteenth century did to the Amerindians.

Those Andromeda strains may be waiting silently and secretly in the liquid-nitrogen canisters of military laboratories in Maryland, Baghdad, Tel Aviv, Beijing, Johannesburg, or some place we do not even imagine. They are microbes created by army biological research laboratories in many parts of the world that are more infectious, more lethal, more intractable than anything humans have ever experienced. The military are the custodians of the biological agents that may do us all in, that may finish the job that the conquistadors started with the other Americans nearly 500 years ago. It is those "unnatural," human-designed pathogens that should scare the hell out of everyone.

Having vented but not relieved the anxiety over some army's superbacteria or -viruses, let us return to the chapter's business of health and disease during the early post-Columbian years. From that time onward when we speak of the American disease, it is not yellow fever or cholera or Chagas' or some New World virus; it is malaria. Malaria is as American as the heart attack and apple pie and has or had been entrenched from New York to California to Argentina. There are no questions as to where malaria is or was in the Americas; the mystery is when and how it got here.

There is a long-standing belief that Mother Nature is humanity's friendly druggist; wherever there is a regional illness, she has placed a specific plant remedy for it. Working from this botanical drug–disease proposition, a medical historian can reason backward; if the plant is or was there, then the disease must have been there from the beginning as the natural order of the symmetry of life in that locality.

Malaria's origins in the Americas has long been debated among malariologists. One side of the debate is adamant that all three human malaria parasites in the Americas are imported, that there simply was no malaria in any place in North, Central, or South America before the coming of the Europeans and their African slaves. On the other side of the debate are advocates of a pre-Columbian malaria. Their major piece of evidence is South America's first and foremost antimalarial, quinine, a drug produced from the bark of the cinchona tree.

The Aztecs called it *quina-quina.* They told the early Jesuit missionaries that it was used to cure illnesses, particularly febrile illnesses. To the Jesuits who came from malarious Spain, fever essentially meant the tertian fever of malaria, and they tried quina-quina for their attacks of the ague.[6] It had a miraculous effect and in 1632 a Father Alonso Messias Venegas brought back the bark to Rome, a highly malarious city. For more than three hundred years quinine was the one and only antimalarial drug, curing princes and paupers, popes and prime ministers. Even as I write, physicians throughout the tropics are reaching for the bottle of quinine tablets to treat their patients acutely ill with multidrug-resistant malignant tertian *(Plasmodium falciparum)* malaria.

If the quinine alkaloids of the cinchona bark were narrowly specific for malaria, it would lend support to the belief that some Great Wisdom put the tree there to benefit the malarious indians, but it is not. Quinine, like most botanical

6. It may be questioned that if the malaria importation theorists are right, then where did the Jesuit missionaries get their malaria if not in the Americas? The answer would be they acquired the malaria in Spain where most of the infections would be of the benign tertian type caused by *Plasmodium vivax.* This malaria parasite has a long life in its host, up to at least five years, during which time there are repeated attacks as the parasites emerge from their dormant phase in the liver. And although *Plasmodium malariae,* the cause of quartan malaria, has no persisting liver stage, it can circulate as a low-grade infection for many years; one such infection of 52 years has been reported.

drugs, has a spectrum of diverse activity. A quinine derivative, quinidine, is an important drug used in treating an irregularly beating heart. Quinidine, in its turn, is also active against the malaria parasite, and tropical doctors without quinine—it has been in short supply—are advised to give their malaria patients the more readily available quinidine.

In the same context, our most modern drug for falciparum malaria, almost the last string to the therapeutic bow, is the Chinese botanical Qinghaosu from the sweet wormwood *(Artemesia annua).* It may be a very modern drug, but it was used as a medicine some 2,000 years ago. That might indicate a Chinese malaria problem during the Han dynasty. But when Ge Hang wrote of it in 340 A.D. in his *The Handbook for Emergency Treatments,* it was not for use against a tertian fever but rather to treat hemorrhoids. The Chinese may, in fact, never have appreciated the antimalarial properties of Qinghaosu. In 1693 the Kiangxi emperor was stricken with malaria. His court physicians were unable cure him; either they didn't have or didn't know about Quinghaosu. The French missionaries to China at that time always carried quinine with them, and one of those missionaries administered it to the feverish sovereign. The Taiwanese art historian Chang Lin-sheng writes (*Orientations,* October 1995: 52), "The emperor was cured, and from then on gave the French missionaries his utmost trust."

And what should we make of the recent report of the antimalarial activity of cocaine?

What happens when we examine the scrappy written records of the ancient Amerindian medical codexes and the more voluminous accounts by the early missionaries and explorers of the Americas?

It is not absolutely required to confirm the presence of malaria under the microscope, to discern the stained parasites within the red blood cells. Those who have had malaria know that it is a singular experience unlike any other infectious illness. The onset with its tooth-rattling rigor (in Papua New Guinea pidgin it is called the *guria,* the earthquake) gives way to drenching sweats. The great fever breaks, but the respite is short and the rigor-sweats cycle begins again. The cycle repeats like clockwork every 48 hours.[7] These tertian and quartan fevers were well recognized by the ancient physicians as far back as Hippocrates (c. 460–c. 370 B.C.) as a distinct disease entity. Not until the mid-eighteenth century did it become known as malaria, or mal'aria.

So the Spanish and other early European explorers, missionaries, and colonizers were aware of malaria since it was so common in their own countries. The Portuguese, who came to occupy Brazil, had imported so many African slaves into Portugal by the late fifteenth century that their falciparum malaria ignited a series of epidemics so intense that the Tagus valley was almost depopulated. However, there is nothing in the Maya, Aztec, Olmec, or Inca records that indicates malaria. It has also been suggested that these great American civilizations could not have thrived in a malarious setting. Of course, Rome was malarious, but the "day" that Rome was built might not have been a malarious one. Nor do any of the first Europeans in the New World mention malarial fevers. Cortés in his reports to King Charles V told him of how healthy everyone was and what a wonderful time

7. The 48-hour fever cycle of *Plasmodium vivax* and *Plasmodium falciparum* is called tertian fever whereas the 72-hour cycle of *Plasmodium malariae* is called quartan fever because it is calculated in the old Roman way in which the onset is time 1 rather than time 0. It is like the Chinese way of birthdays—you are born 1 year old rather than starting out at 0.

Also, the regular tertian and quartan cyclical fevers is the hallmark of malaria. However, as the untreated infection progresses, it may assume many other confusing guises, such as dysentery and pneumonia. Clinically, malaria has been called the great mimic.

he was having beating up on the Aztecs. So, from the record there was no American malaria before Columbus.

Living relics also open the past. The human heirloom of malaria is in the blood. Wherever malaria has been entrenched for thousands of years, selective evolution has sieved the humans with mutations that rendered them resistant or partially resistant. These mutations all, in some way, involve the red blood cell. The best known is the sickle cell trait in West African blacks (and African Americans of West African descent) which allows children with this trait to survive infections of *Plasmodium falciparum.* Another mutation is Duffy factor negativity of the red blood cell membrane which makes West Africans totally insusceptible to *Plasmodium vivax.* There are other protective blood mutations—ovalocytosis in Melanesians, G-6PD deficiency in Mediterraneans. It would take a book to tell of the blood and malaria; suffice it to say where malaria has been a long companion of a people, there is funny blood. No Amerindian in South America or North America has any blood mutation that bespeaks of ancient malaria. So, the relics too deny New World malaria before the arrival of the Europeans and the Africans.

I had concluded that there was no malaria in the Americas before 1492 when my brother-in-law, the malariologist, called me up. He had been reading the draft manuscript, and although essentially an experimentalist, he too is getting to an age when history becomes of interest. He is also the most tenacious person. He had been having serious thoughts about the origin(s) of the malarias in the Western Hemisphere and wanted to tell me about the new research of Dr. Tom McCutcheon.

Old timers like myself can look down a microscope and tell from the stained film what malarial parasite is present in the blood. After twenty or thirty years we become pretty good at making the correct diagnosis. Now come the molecular-oriented malariologists who tell us that there is more to it than meets our eye. Their gene-sequencing analyses, yielding RNA and DNA homologies, show that *Plasmodium vivax* is a composite of types or subspecies. That's OK. We old timers long suspected that this was the case. *Plasmodium vivax* from the different parts of the world acted differently: the strains from cold climates, such as from Russia, have long periods either before first spawning from the liver or between relapses; other types from hot climates such as Southeast Asia have short periods. There also are clinical differences and drug-response differences between geographically distant isolates. But what excited my brother-in-law vis-à-vis this chapter was McCutcheon's molecular characterization of a malarial parasite of South American monkeys, *Plasmodium simium.*

Under my microscope *Plasmodium simium* looks exactly like *Plasmodium vivax.* It is now known that South American monkeys such as the howlers and owl monkeys are susceptible to infection with *Plasmodium vivax* and can be useful experimental animals for the study of that parasite. Unlike *Plasmodium falciparum,* no technique is available for the continuous, test tube cultivation of *Plasmodium vivax.* This susceptibility and experimental cross-infections from monkeys to humans points to humans as the original source of the monkey infection. And when it got into the monkeys, the biologists called it *Plasmodium simium,* but it really and truly was *Plasmodium vivax* that had come into South America with the Spaniards. What Tom McCutcheon now has revealed by his elegant sequencing of the parasite's RNA is that the monkeys don't have Spanish *vivax* but that what is called *Plasmodium simium* is *Southeast Asian* vivax!

This astonishing fact implies that when northern Asians came through the Bering Straits, they brought the malaria parasite with them and it survived long enough for the immigrants to disperse to the warmer, malaria-friendly climate of California or, when southeast Asians came, swept by misadventure into irresistible currents, as we believe happened to the Japanese in coming to Ecuador, they brought benign tertian malaria to the tropical New World.[8] However, whatever the route and whoever the Asians, if *Plasmodium simium* is an heirloom relic, then benign tertian malaria probably came to the Western Hemisphere before Columbus did.

Then before I could leave this topic I came across a forgotten paper that I had missed in researching the literature for this chapter. It was another neotropical malaria relic report. In 1973, Raul Cantella and Alejandro Colichon of the Universidad Peruana Cayetano Heredia in Lima, Peru, went to a remote tributary of the Amazon to study an isolated nomadic group of Amerindian hunter-gatherers known as Campas. As part of the study they made blood films, and these were sent to Alexander J. Sulzer at the U.S. Public Health Service's Centers for Disease Control in Atlanta. When Sulzer and his colleagues Neva Gleason and K. Walls stained and examined the films for malaria parasites, they found that 83 percent of the Amerindians were positive for *Plasmodium malariae*—possibly the highest infection rate of this parasite ever recorded. This was perplexing. How did this supposedly isolated group with little or no previous contact with outsiders come to get so much malaria? Where did

8. *Plasmodium vivax* is present as far north as Korea where it was an important medical problem for U.S. troops during the war there. I have been told that in Korea vivax malaria was mainly a venereal problem for the military. Antimalarial discipline was considered to be good, but exposure to the mosquitoes' attacks occurred when the soldiers visited the bordellos.

it come from? Monkeys? Like *Plasmodium vivax, Plasmodium malariae* infected the local South American monkeys (and underwent a name change to *Plasmodium brasilianum* in the process). It would thus seem that both these human malarial parasites were exchanging back and forth between monkey and man in the jungle. All well and good, but where did the original *Plasmodium malariae* infection in man or monkey come from?

Like all malariologists, Sulzer and his colleagues were interested in the antiquity of malaria in the New World. Was the *Plasmodium malariae* in the Campas Amerindians a true relic from a pre-Columbian time or a leftover from a much later time, from some fleeting contact with an infected (via a mosquito, of course) person from the outside? If a true relic, where did it come from?

Pre-Columbian quartan malaria poses a different set of problems and a different set of historical implications than the Asian *Plasmodium vivax* because *Plasmodium malariae* is considered to have its origins in Africa as a parasite of chimpanzees. If we believe that the *Plasmodium malariae* infections in the isolated Campas Indians stem from Africa, then it follows that West Africans came to the Americas hundreds if not thousands of years before Columbus; this would make these first African Americans as native as Native Americans

Some evidence does indicate that small communities of Africans were scattered throughout eastern regions of the Americas by the time of Columbus's voyages. There were numerous "black sightings" by the early explorers such as that made by Vasco Núñez de Balboa on his march to the Pacific. Yam and taro, African food crops, were in South America long before 1492.

There is the mystery of the dog that didn't bark. Columbus commented on the silent dogs of Cuba, like the quiet basenji of tropical Africa. A real curiosity is the possibility of pre-Columbian Africans in Alabama. Early mound makers lived there, and at one site, known as the Rodean Mounds, archaeologists found money cowries *(Cypraea moneta)*. These common little cowries are from tropical Pacific seas, but Arab slave traders brought great loads of them to West Africa where they were used as currency. M. D. W. Jefferys in his article "Pre-Columbian Negroes in America" (*Scientia* 88 (1953): 202–18) argues that this is yet another proof that the African American is a native American.

The train of malaria-connection logic for an early African American goes a bit off track with the absence of any evidence of pre-Columbian, preslavery *Plasmodium falciparum*. This is *the* common malaria parasite of Africa, and there were highly efficient American anophelines capable of acting as its vector awaiting its introduction. Perhaps it was brought in with the Africans storm tossed onto American shores and decimated the immunologically defenseless, small-group Indian tribes leaving only the smoldering *Plasmodium malariae* as a souvenir of those first immigrants from Africa. This is all diverting, amusing historical-epidemiological speculation. The hard facts are that after slavery, malaria and so many other diseases, brought to the Americas with virtually each cargo of Africans in wretched bondage, found a permanent residence in the Western Hemisphere.

Chapter 5

Coming to America 1638 to 1865: From Africa on the Slave Ship Named *Desire*

KWAME THREW A handful of pebbles into the bulrush millet and a flurry of protesting quelea finches rose into the air. They soon flew to another patch of the tall, tasseled plants where Kwame's cousin, Kwisi, waited with his pile of pebbles. The field was ringed by finch-chasing boys to protect the ripening crop from the marauding quelea birds. The rains had been bountiful and the crops were lush; the people of Kwame's village would not hunger during the coming dry season. Some of the grain would be made into milky-white bush beer drunk on the nights when the moon was full and the people danced. The proud-breasted young women, shining with palm oil, arms reddened with henna, faces whitened with ash pigment, would stamp and sway in their joined circle. And as they danced, they sang of love and its fulfillment. Then the young, unmarried men would advance as the drummers sped the rhythm, and they would leap and cavort in the wild choreography of teenage youth. In the shadow of the fire Kwame and his companions would imitate the dance of their elder brothers.

Kwame threw another handful of pebbles and the queleas flew. Soon, thought Kwame, he would join the dance of men. Soon he would take a woman. Not long ago he experienced the telltale sign that he was coming of age, that he was sexually mature; his urine was tinged red with blood. It was like his sister who produced the sign of the blood when she was ready to be given away in marriage. She was but a woman and bled only once in a moon whereas he was a man and he urinated red all the time. Men were indeed superior to women.[9] Yes, soon he would dance the dance of men, his penis adorned with a feather-decorated sheath that would swing seductively to the beat of the drum.

Kwame threw another handful of pebbles into the millet. The day was waning and the contest between bird and boy was coming to a temporary end. Kwisi joined him, and the two walked the track to their village and the welcome of the evening meal.

His father lay lifeless on the dancing ground, the blood congealing in the dirt. His mother cowered by the family's hut, screaming, shaking, bound hand and foot with heavy bush rope. His sister, similarly bound, trembled uncontrollably next to his mother. It was then that the astounded Kwame was seized by a giant of a man as dark of skin as himself. Kwame's immediate thought was that the village had been attacked by the neighboring tribe a day's march to the north. Traditional enemies, the two tribal groups had been raiding each other's villages, as the griots sang in the oral history of the tribe, "from the beginning of time." In the idle time, before the planting season of the early rains, Kwame's tribe had raided their enemy's village and brought back prisoners—two men, a woman, and a child about his age. There

9. The blood in Kwame's urine, the hematuria, was caused by the parasitic blood fluke, *Schistosoma hematobium,* living in the veins draining the bladder. The eggs of the worm cross from the vein to the bladder where they cause small hemhorrages and inflammation. It is a disease transmitted by a freshwater snail and so common in many parts of Africa that the hematuria in young men is looked on as male menstruation.

was great excitement when the captives had been paraded through the village and squabbles when each family indicated the portion it wished to have. It had been a stupendously satisfying feast, particularly so because bush meat had been in short supply during the dry season, and at any rate even a bush buck was never as tender or as tasty as human flesh. Now, as his heart contracted in fear and his mind numbed, his last thought before he passed into oblivion was that it was his turn to be the "long pig," the makings of a feast.

The sharp bite of thc whip brought Kwame to consciousness. The man who had flogged him stood menacingly above; a person of strange feature and dress—swarthy but not black skinned, beaky thin nose, thin lips, clothed in a long, white body-enveloping robe, a cloth draped over his head and shoulders. This man of alien appearance shouted orders to his black minions who bound the villagers and flogged them to make them stand. An iron ring was put around each captive's neck and from this he or she was bound by a chain to a long wooden pole. Prodded and beaten by the Arab slaver and his Africans, the stumbling line of yoked captives was forced to march from the village. Kwame painfully turned his head for one last sight and was astounded see his chief sitting in front of his hut, besotted with drink, a heap of bright cloth and trinkets before him, surrounded by his laughing, drunken sycophants and favorite wives.[10]

The next weeks were filled with indescribable misery. Never relieved from their yoke, the captives marched the long miles each day, broiled under the African sun and sus-

10. It is doubtful whether the slave trade could have existed or persisted to such a horrendous dimension without the Arab slavers and the connivance of the African tribal chiefs. H. H. Scott in his monumental *History of Tropical Medicine* (Baltimore, Md.: Williams and Wilkins, 1939) comments, "So long as any part of the African coast was under Mohammedan control there was slave-trading and slave raiding in the hinterland."

tained by a meager single meal of grain and cassava washed down with a gulp of water. Each day's route was marked by the rotting bodies and skeletons of those who had gone before. Some of Kwame's group contributed fresh bodies as grisly signposts. Some died of dysentery. Some too weakened to walk were bludgeoned to death and their bodies hacked away from the chain. One night hyenas attacked a fitfully sleeping woman; her screams aroused the others and the slave caravan guards, but by the time the animals were chased off, her legs had been gnawed and bitten to the bone. She was quickly dispatched by a captor's cudgel.

Finally they came to the sea. Fewer than half had survived the long weeks' march. Some had died of explosive diarrhea. Some of the children had high fevers, became comatose, and were slaughtered when they could not be roused.[11] Some, in the depths of despair, refused to eat and wasted away. They, too, were killed when they became too weak to walk farther. The survivors were crammed into a prisonlike house, a barracoon, where they were chained to iron rings set into the stone walls. Here they stayed for several weeks. More died. When Kwame peered through the narrow slit that served as a ventilating window, he saw the frightening expanse of the sea extending to nothingness beyond the horizon. He saw great masted ships flying brightly colored cloths, some decorated with a red and blue cross, some with red, white, and blue bars and a circlet of stars in one corner. He saw large canoes carrying tribal chiefs and dignitaries laden with cloth and rifles and trinkets of many sorts that seemed to be gifts from the swarthy men and the

11. Their deaths were probably due to cerebral malaria although bacterial and viral meningitis would also have taken their toll in the children whose immune systems had been weakened by the stress and deprivations of the slaves' long march to the sea.

white men who were the masters of the great ships and this terrible place. He saw other captives bound to posts set in the ground leading to the sea. He saw emaciated men and women lying on the ground, as if dead, their eyes staring in unseeing oblivion.[12]

Slavery, the importation of Africans to the New World, began a short 11 years after Columbus came to the Caribbean. The Spaniards bought and brought slaves from West Africa to work the mines in their new dominions of Mexico, Peru, and Panama. They needed African slaves because the Amerindians were dying in wholesale lots from disease and overwork. One bishop, Las Casas, sent by the Spanish to make sure that the Amerindians they were killing died as Catholics was horrified by this inhumane treatment. He petitioned the king, Charles V, to allow the importation of African slaves to relieve his enslaved parishioners. The king was only too happy to accede to his request, and the floodgate to the African slave trade was flung open.

For the first hundred post-Columbian years or so slavery was pretty much a Catholic monopoly. In 1493, "God" in the guise of a Papal Bull divided the non-European world between Portugal and Spain. To Portugal went Africa and all its souls; to Spain went the New World and all its souls. A year later Portugal and Spain confirmed the Pope's division by the Treaty of Tordesillas, but now Portugal got Brazil because no one was quite sure where Brazil was in 1494. Business was good for both parties: the Portuguese collected

12. These were the victims in the last stages of West African (Gambian) sleeping sickness. This is a disease caused by the trypanosome, a protozoan pathogen transmitted by blood-sucking tsetse flies. In the "sleeping" comatose people seen by Kwame, the organisms had gone from the blood stream to invade the brain and central nervous system. The causative organism wasn't discovered until 1901. During the slave trade years the disease was called the "Negro distemper."

slaves and made money; the Spaniards bought the slaves and made their fortune from the forced labor. Even the "petty" Catholic nations prospered from the butchery in Africa. Charles V gave a Flemish merchant exclusive rights to sell him 4,000 Africans a year. The French were big-time slave dealers and made money—a lot of money. Nantes became a rich city as a center for the fleet of slave ships based there. The French were still slaving in the nineteenth century. By then the Nantes slavers had sent some 1,800 expeditions to Africa where they collected approximately 500,000 pieces of merchandise. They sold this human cargo mainly in Haiti but later on the Louisiana coast.

The English could never quite come to a firm resolve on slavery. It was so abhorrent. It was so profitable. Queen Elizabeth I said that the slave trade was a "detestable act which would call down the vengeance of heaven upon the undertakers." She then knighted John Hawkins, the master slaver who commanded a fleet of ships trading in the humans he took from West Africa to the West Indies. The flagship of his fleet was named *Jesus*. Finally, in 1808, Parliament passed a law prohibiting slavery. In 1824 it made it a felony and declared that trading in slaves was an act of piracy. In 1817 the British government gave the Spanish government £400,000 and the Portuguese government £900,000 to persuade them to sign treaties outlawing slavery and permitting British warships to interdict their slave ships. Money mostly down the drain. The Spanish and Portuguese carried on almost as before, adopting a new ploy—their slave ships flew the U.S. flag until they were beyond the British squadrons stationed off the coast! It was this flag, the stars and stripes, that Kwame saw as he descended into the hell of the middle passage.

Kwame was a fortunate survivor of the middle passage.[13] He came on the slave ship named *Desire*. He was sold in a South Carolina market. When he was taken to the plantation, he was overcome with great fear and his bowels loosened. In the feces so unceremoniously dumped on the South Carolina soil there was that other African, the egg of the hookworm *Necator americanus*. In the warm southern sun and moist soil the egg embryonated and the larva hatched to await penetrating the next barefoot passerby, black or white. Hookworm had come to North America. That evening an American mosquito, *Anopheles maculipennis*, fed on Kwame's blood, blood that contained the sexual stages of the malaria parasite *Plasmodium falciparum* infective for the mosquito. Twelve days later that mosquito would feed again, this time on the plantation owner.

We do not know just when or where these and other African diseases came into the Americas. However, this fictionalized account echoes expert opinion that "the negro was a hive of dangerous germs." The slaves brought leprosy, filariasis (elephantiasis), hookworm, yaws, and yellow fever into the United States. Malaria in all the Americas and schistosomiasis in the Caribbean and South America can speculatively be added to that list. However, malaria was the most constantly pervasive of these diseases, and it was malaria that, paradoxically, was so important in the perpetuation of slavery.

Tobacco, reviled today as the addictive enslaver of the pulmonary tract, was also a literal enslaver. It was nicotine

13. It was a time of indifferent cruelty and not only to the Africans. Perhaps the most vivid description of a middle passage is not that of a slave ship but of a British hospital ship of 1739! "The men were put up between decks in small vessels where they had not room to sit upright; they lowed in filth; myriads of maggots were hatched in the putrefaction of their sores."

that brought Africans to labor on the Virginia plantations. Between 1680 and 1786 the British forcibly brought 2.1 million Africans into their American colonies.[14] At first they were taken from Gambia, Ghana, and Sierra Leone, but as the human merchandise became scarce in that region, the slavers extended their sweep to Dahomey, Benin, Angola, and Zambesi. Of the millions taken, an estimated half-million died each year. It was like a deforestation of humans.

As in all times of great cruelty there were those who were revulsed by the dehumanization and slaughter of the Africans. In time those "liberals" would probably have prevailed and slavery would have been abandoned in the United States—the slavery not only of blacks but also of whites. Thousands of Europeans were indentured to meet what seemed to be the insatiable demand for the labor needed by the southern plantations. The indentured whites also had a terrible time and lived under appalling conditions. But whereas white slavery gradually diminished, the Africans' unique evolutionary adaptations to resist the pernicious effects of malaria kept them in bondage.

Plasmodium vivax is called the benign tertian malaria parasite because it doesn't kill, but oh, how prolongedly sick and weak it makes you feel. Untreated falciparum malaria, the malignant tertian malaria, kills. I've had both and, frankly, felt much worse with my "benign" attack.[15] Thus in the New

14. The Africans were mostly taken to and sold in the plantation South, but the North was not without slavery. The Dutch landed their first cargo of slaves in New York in 1619. By 1770, approximately 150 slave ships were running out of Rhode Island.

15. As an "economic" pathogen *Plasmodium vivax* is the more important parasite because of its repeated attacks over a long period of time (arising from dormant parasites in the liver) and its ability to produce a chronic anemia. It is an unproductive laborer who is repeatedly feverish, aching, and anemic.

World agriculture the southern planter and the indentured white slaves were being knocked over, like so many ninepins, by malaria. But the Africans kept working. They were totally, solidly, and naturally immune to infection with *Plasmodium vivax* and partially immune to falciparum malaria.[16] They were a source of labor that was too cheap and too healthy to let go.

Meanwhile the ecological changes wrought by agriculture were further intensifying malaria endemicity. The introduction of rice farming into the southern South created thousands of acres of wetland ideally suited to malaria-carrying mosquitoes. The malaria problem of the northern South was one of seasonality. In tobacco growing areas like Virginia the winters were too cold for the transmission of malaria through the mosquito. But come spring and planting time, the mosquito and malaria were back. Only the immune black could carry on.[17] One is reminded of the plight of the Indian farmer

16. I have explained the mechanisms of natural resistance to each malaria parasite in my earlier book *New Guinea Tapeworms and Jewish Grandmothers*. Briefly, the red blood cells of West Africans and, of course, their descendants lack the essential molecular receptor on their outer surface to allow *Plasmodium vivax* to "dock" and invade, and thus are totally immune to this parasite. Africans are also partially immune to *Plasmodium falciparum* because many carry the sickle cell trait hemoglobin that inhibits the malaria parasite's growth. Unfortunately, the double dose of the sickle cell gene leads to early death—an evolutionary trade-off between this fatal condition and the even more fatal falciparum malaria. The noted malariologist Dr. Louis Miller has hypothesized that the tendency to high blood pressure and iron overload present in African Americans are also genetic heirlooms of malaria-resisting traits.

17. A price must be paid for the relief of seasonal malaria. Despite many years of research there is still no exact understanding of the immunological mechanisms by which humans or animals acquire clinical and parasitological resistance to malaria. Observation, however, has shown that in some respects more malaria is better; constant exposure (immune stimulation) takes its toll of the young, but the survivors maintain a reasonably good immune resistance throughout life. The dry-season, or winter, respite

in Rajisthan in 1994 when unusually heavy rains brought a lush crop of wheat but also a proliferation of malaria-transmitting mosquitoes. There was so much sickness and so many deaths from malaria that the abundant harvest could not be realized.

To colonial America's white planters this difference in susceptibility to malaria proved the superiority, the delicate refinement of the whites over the blacks. The black wasn't sufficiently advanced as a "species" to acquire such an exquisitely "human" disease. The medical philosophy of the time supported such outrageous notions. It was a time before the discovery of microbes as agents of infectious disease. Medicine of the seventeenth and eighteenth centuries hadn't advanced all that much beyond the imbalanced-humor beliefs of the ancient Greeks. American and European medical theory on the causation of sickness was still very much yin-yang. Etiology was considered in terms of imbalance—insufficiency of nervous tensions, overexcitation of body organs, too much bile, too much blood. Phlebotomy, the letting of blood, to put the patient back to normal was a popular catchall treatment, as common as the broad-spectrum antibiotic therapy of today.

Perhaps the slaves might have been better served by not having a physician in attendance. The medical practices, such as phlebotomy, and remedies, such as mercury and antimony compounds, could do more harm than good to a sick body naturally trying to restore itself. By twentieth-century standards the seventeenth- and eighteenth-century physicians could only be judged as incompetent if not outright dangerous ignoramuses. There were no American medical schools

of seasonal malaria does not induce this level of immunity, and there is a good deal of morbidity and mortality with the onset of the malaria season.

until 1775 when one was established at King's College (later Columbia University). Most of the doctors became doctors by apprenticeship, a long seven-year stint starting at 13 or 14 years of age.

Then there were those curious physician-preacher hybrids. They were mostly nonconformist pastors who could find no paying ministry. During their theological training they would attend lectures in anatomy and what was then known of physiology. At first these preacher-doctors were English who emigrated to America but later Harvard picked up on this joint degree. They had no practical training or experience when they were let loose on the public and picked up what knowledge and skills they could on the job, over the years, saving bodies and souls. In light of today's soaring medical costs, it might not be a bad idea to take another look at that earlier educational system. Maybe our medical schools should recruit their students from theological seminaries, convents, yeshivas, and so on, giving special priority to candidates who have taken vows of poverty—Carmelites, Benedictine monks, sadhus, and so on.

The slave owner had to keep his black chattels healthy and working. If they were not quite human, should they be treated by someone like a veterinarian who specialized in that species of domestic animal? Should they be given the same therapies as whites? Even the sovereign remedy of bloodletting was in doubt. Thomas Jefferson, a famous waffler on the issue of slavery (except for his personal sexual satisfaction), was adamant on this issue. "Never bleed a negro," he says in his garden book.[18]

18. Jefferson might have been right albeit for the wrong reasons. The unshod black slaves living and working in unsanitary conditions undoubtedly had the high hookworm load that causes chronic anemia that "therapeutic" bloodletting would only aggravate.

Thus, the Africans gave the Americans the classically tropical diseases of malaria, filariasis, hookworm, schistosomiasis, and yellow fever. In good return, the Americans gave the Africans the tropical disease syphilis.

Syphilis had spread rapidly throughout Europe after being imported from the tropical Americas. Columbus's crew may well have been the first importers. As early as 1500 European physicians recognized the congenital type of the disease passed from mother to fetus. Later, from 1876 to 1885, the English physician, J. Hutchinson was to extend and codify these early observations, particularly in his description of the typical peglike malformation of the front incisor teeth present in about one-third of the cases of congenital syphilis.

In 1992 Keith P. Jacobi of Indiana University's Department of Anthropology and his colleagues decided that in addition to the general poor health of the slaves there was, specifically, rampant congenital syphilis. They estimated that a congenital syphilis rate of approximately 50 percent had been the chief cause of the horrendous infant mortality, as high as 50 percent for Caribbean slave populations.

Jacobi's group then went on to examine the dental remains from other cemetaries of the slave era. Was syphilis a Caribbean problem or was it more widespread? In the exhumations from a slave cemetary in Cedar Grove, Arkansas, they found Hutchinson's teeth; syphilis had affected the slaves of Arkansas. However, in the remains from a cemetary of the First African Baptist Church of Philadelphia, a congregation of free blacks dating from 1800, there were no dental signs of syphilis. There were no signs of ill health or high childhood mortality. Their condition and life expectancy were estimated to be similar to the whites of that time.

Once freed, some blacks pursued the American dream

and became wealthy entrepreneurs. One, Paul Cuffe, made a return visit to Sierra Leone, that steamy, seedy British colony of Graham Greene's *The Heart of the Matter*. Cuffe, despite his own personal success, believed that America could never be home to the blacks and agitated for the return to Africa of all freed slaves. In 1815, with the cooperation of a group of British philanthopists, Cuffe arranged for the first black recolonization of Africa with the settlement in Sierra Leone of 38 freed slaves from the United States and Canada. The spirit of return now took fire in the hearts of blacks and in the minds of white abolitionists who founded the American Society for Colonizing Free People of Color of the United States.

However, it was Liberia, not Sierra Leone, where most of the freed slaves ultimately went. It was not exactly a repatriation. Very few of the approximately 15,000 settlers that came to Liberia were of Liberian origin. In the end it fell under a colonial rule by blacks of American origin that was just as "colonial" in character as that imposed by any white European power in Africa. And the black colonists of Africa died with the same great mortality rate as did the white intruders. A major killer of those now-nonimmune immigrants was yellow fever.

I once had a squeamish friend who loved lobster. It was his contention, never put to the test, that if you put the lobster in a pot of cold water and very slowly raised the temperature, one degree at a time the lobster would be unaware that it was being boiled to (a merciful) death. And so it is with chronic diseases. Many chronic diseases are the leading causes of mortality but dying by inches does not strike terror in the collective psyche. It is the swift, killing, acute diseases appearing suddenly that cause terror and panic. The bubonic plague has long been the metaphor for such microbial terroristic threatening but from 1649 to 1905 that metaphor in the Americas, North and South, was yellow fever.

Yellow fever is caused by a virus; more specifically, it is a flavivirus of spherical shape whose genetic constitution (its genome) is RNA rather than DNA. Most importantly, it is an arthropod-borne virus, an arbovirus, transmitted by the bite of an infected female mosquito. Its ancestral home is tropical Africa where it infected (and still infects) humans and monkeys. All primates are susceptible to infection with the yellow fever virus. For some eccentric reason of host specificity, the hedgehog is also highly susceptible and usually dies if infected.

Yellow fever was recognized as a disease entity by seventeenth-century physicians, some 300 years before the discoveries of the causative virus and its transmission by the mosquito. Although those early clinical descriptions were remarkably accurate considering the state of medical science at that time, there was some confusion in delineating it from an acute, usually fatal, form of malaria, blackwater fever, in which the massive breakdown of red blood cells is revealed by the borscht-red color of the urine.

In the eighteenth century there were no peer-reviewed journals to adjudicate medical disputes and sometimes scientific partisanship became violent. In 1740 a Dr. John Williams who had been a physician-surgeon on a slave ship plying between the west coast of Africa and the West Indies declared, in Jamaica, that blackwater fever (then also called remittent bilious fever) was a different disease entity than yellow fever (then also known as hemogastric pestilence). Dr. Parker Bennett, also of Jamaica, declared Williams a charlatan in this belief; the bilious remittent fever and the epigastric pestilence were one and the same. Words were exchanged, then intermediaries. They fought a duel. Both were killed.

However, both Bennett and Williams were generally accurate in their description of yellow fever: the early symptoms of fever, flushed face, headache, joint and muscle ache; a flu- or malaria-like illness that lasts for about three days (Williams was correct, blackwater fever *is* different from yellow fever). Often the disease abates after this brief illness; in some cases "yellow fever" ends here, whereas in other cases it is only a deceptive remission and by the seventh day the disease begins to rage with renewed savagery. Now there is intense epigastric pain with vomiting of bile and blood, the skin colors the deep yellow of the acutely jaundiced, the heart beats irregularly, and the patient becomes delirious and then comatose. Miraculously, even those in the late stage of the disease may survive, the estimate being that "only" 50 percent of the delirious-jaundiced die.

The devil's agent of yellow fever is the aedine mosquito which transmits the virus from person to person, or monkey to person, or monkey to monkey. There are numerous species of *Aedes* mosquitoes; they are an adaptable family with member species flourishing from sub-Arctic to tropical environments. They are rather handsome little creatures with black bodies sprinkled with white spots. They like to feed under conditions of low light, in the shade, on cloudy afternoons, and at dusk—crepuscular biters. The aedines of medical importance breed in small collectors of clean water such as rainwater collecting jars, discarded tires, and derelict tin cans. We once owned a house in a Hawaii that had a high attack rate of Aedes in the late afternoon. It was impossible to sit out on the lanai for the sundowner glass of wine. In desperation my wife called the health department which sent a sanitarian to inspect our premises. He showed my wife all the pots I stored under the lanai, awaiting the orchids I would put in them, and each with a colony of Aedes larvae swimming in the collected water. "Lady," the inspector said, "you should tell your husband where all these mosquitoes are coming from." To which my wife tells me she replied, "My husband is Professor of Tropical Medicine at the University and he lectures on medical entomology." Well, my father was an electrical contractor and our light switches never really worked right.

It was this ability to breed in small, confined collections of freshwater that brought the prime transmitter of yellow fever, the *Aedes aegypti* mosquito, from Africa to the Americas. The life span of an adult mosquito is brief, one to two months, too short to survive the long transatlantic voyage of the slave and cargo ships under sail. But the numerous barrels of freshwater aboard those ships ensured the succession of generations of the Aedes. Down below, the crew and slaves ensured a plentiful blood supply for the female mosquitoes, and the rotting fruits and vegetables in the larder gave food for their strictly vegetarian male consorts (female aedines can also live on plant juices but need blood to lay their eggs; they also live longer on blood than on fruits and vegetables).

The virus could have come from Africa by human and by mosquito. It could have come in the blood of slaves who were asymptomatic, semi-immune carriers as well as of sick sailors.[19] It could have come in the *Aedes aegypti* themselves

19. Yellow fever was the most greatly feared disease of sailors of that period. Entire crews might be affected during those long voyages and the mortality might be 70 to 100 percent. The ghostly *Flying Dutchman* was such a ship. First cursed by murder and then fatal yellow fever, it was condemned to haunt the seas around the Cape of Good Hope. And if you looked very carefully, you might see an *Aedes aegypti* mosquito hanging, together with the dead albatross, around the neck of the ancient Mariner. Coleridge's "Rime" is believed to be the story of a yellow-fever-struck ship.

who fed successively, from generation to generation on infected slaves and sailors. Also, modern research has shown that the mother mosquito can pass the yellow virus to her eggs. The eggs are not adversely affected by the virus and when they hatch into larva and pupate the newly emerged adult is already infected and, if a female, capable of transmitting the infection to human or monkey at its first blood meal.

Sometime in the sixteenth or early seventeenth century *Aedes aegypti* came to the West Indian outposts of France, Spain, and England. Shortly thereafter they invaded southern Mexico. Their numbers increased but still there was no yellow fever. Then in 1647–1648 an unknown concatenation of conditions led to explosive simultaneous outbreaks in Havana, Barbados, Guadeloupe, St. Christopher, and Mexico's Yucatan peninsula.

For many of the highly susceptible Carib and other Amerindian groups, already ravaged by Spanish brutality and "white man's diseases," that first yellow fever epidemic was the final epidemic. The white colonists-planters, commercial exploiters, administrators, and military also vomited the black bile and died. The children, mercifully, were mostly spared, as the infection generally took a milder course in the young. The blacks, who had acquired their immunity early in life in Africa, rarely died of yellow fever. Their survival gave further proof of the economic advantage of their enslavement over that of the Amerindians and indentured whites.

The frontier of the *Aedes aegypti* and the yellow fever virus it held continued to advance. By the end of the eighteenth century it had become entrenched from Brazil to Boston. In the rain forests of the Amazon the virus was taken up by the native species of mosquitoes habituating the jungle canopy. The main source of the blood meals of these aedine canopy fliers were the New World monkeys—howlers, marmosets, owl monkeys, and lion monkeys. They were highly susceptible, and from the eighteenth century onward yellow fever caused great epidemic die-offs of these monkeys.

As European colonization of the tropical Americas progressed, villages, towns, and cities arose at the forest border. At this forest border, the ecotone, the epidemiological-ecological transactions between virus, vector, monkey reservoir, and human were reminiscent of Africa: monkey to monkey by mosquito species of the forest canopy, then monkey to human by aedines of the ecotone, and finally human to human by the urbanized *Aedes aegypti.* But there were no monkeys in Philadelphia when yellow fever struck in 1793 and killed one-tenth of that city's population.

Like a tropical hurricane sweeping from the West Indies, yellow fever made its first North American landfall in Spanish Florida in 1649. For the next 250 years there was epidemic on epidemic in the future United States of America. The southern port cities were particularly hard hit. New Orleans (along with Havana) became the "yellow fever capital" of the Americas. Mobile, Charleston, and Savannah all suffered repeated outbreaks as did the small towns throughout the Mississippi valley from the Gulf coast to, eventually, as far north as Cairo, Illinois. But curiously, for the first 150 years of its presence in North America, this very tropical disease from Africa was most frequently epidemic in the present-day northeastern United States. The lively commerce between the West Indies and northern port cities brought repeated introductions of the virus in sick sailors and *Aedes aegypti* breeding in the freshwater casks of the sailing ships. In this way yellow fever struck New York City in 1702, 1732, 1741, 1743, 1745, and 1747 and yearly from 1793 to 1805. Boston was similarly visited by the virus as were other maritime cities as far north as Halifax, Nova Scotia. And yet, despite the epidemics and deaths in those cities when we think of the story of yellow fever in North America, we think of the Philadelphia story.

Charles Brockden Brown (1771–1810) of Philadelphia is considered by some scholars to have been America's first novelist. Brown begins his novel, a kind of pre-Dickensian tale, *Arthur Mervyn or Memoirs of the Year 1793* with, "The evils of pestilence by which this city has lately been afflicted will probably form an era in its history," and, "The influence of hope and fear, the trials of fortitude and constancy, which took place in this city, have, perhaps, never been exceeded in any age."

It is 1793 and Philadelphia, now a city of about 50,000 inhabitants, is the political and intellectual center of the young United States of America. Late in the summer of that year yellow fever strikes the unsuspecting and unprepared city. No one is spared. That the young and robust are struck down and die within a week makes this pestilence so much more terrifying. The death toll mounts. It is a veritable decimation; by late autumn approximately 5,000 people have died. There is panic in the streets. Brotherly love, "fortitude and constancy" is but a fiction of Charles Brown. Individual acts of kindness and love do occur, but for the most part the survivors abandon the ill and dying. Husbands leave their sick wives, and wives their sick husbands; children abandon their parents, and parents their children. The city is paralyzed; banks shut their doors, newspapers close their presses, civic administration comes to a halt. Philadelphia in the grip of the 1793 yellow fever epidemic is as numbed by terror as London during the bubonic plague.

Finally, with the advent of the cold weather and the disappearance of the *Aedes aegypti*, the epidemic wanes and the city slowly returns to life. The citizens of Philadelphia are thankful for their relief but they remain confused as to its causation and abatement. Their dilemma was that in pursuit of the rational, they could no longer believe in a disease etiology of demons, malicious spirits, or divine retribution for human sinfulness but the explanations of the science of the day also gave no satisfaction. The causative agents of what we now call infectious disease—viruses, bacteria, parasites, and fungi—were unknown and undreamed of. It was to be another 75 years before the chemist of Dijon, Louis Pasteur, was to open that Pandora's box of microbes.

The fallback medical explanations for yellow fever in 1793 were based on observing the obvious. Multiple people sickened and died within a short period of time. The sickness, therefore, must have been passed from one person to another; it was a contagion. But there was also an anticontagionist school one of whom was the extraordinary Dr. Firth. To prove his thesis, he drank the black vomit from a yellow fever victim. When this didn't produce yellow fever, he took the even more heroic step of injecting blood from a yellow fever patient into his arm. Again, nothing happened. He was a very lucky man, was Dr. Firth. The vomit was merely nauseating, but the blood should have contained the virus unless, as was probably the case, the donor was on the way to recovery and no longer viremic.

Another observation: The epidemic(s) began at certain times under certain conditions, when the season turned warm and the humidity was high. During those languid summer months the air frequently held the stench of decay. Therefore, the cause of yellow fever was a miasma, a noxious agent; it was, in a sense, believed to be an environmental disease. The noted Philadelphia physician, Benjamin Rush, was convinced that the miasma of yellow fever emanated from sacks of coffee left to rot on the wharfs. Rush had the prescient insight (which he did not pursue) that somehow mosquitoes might be involved and he noted that in the summer and fall of 1793 "moschetoes were very plentiful."

Still another observation: The onset of the epidemic often occurred after the arrival of an outsider—a ship from another country with foreign crew and passengers, an immigrant influx into town, and even commercial travelers. Therefore, prevention required a *cordon sanitaire,* the denying of entrance of strangers into the community. Thus began the policy of quarantine with all its disastrous economic and social effects.

The total ignorance of the natural history of yellow fever—its causation, transmission, and epidemiology—was to have a crucial impact on the political and economic destiny of the young United States of America. And it wasn't all bad; yellow fever gave us Louisiana and the vast territory acquired by the Louisiana Purchase.

The history of Haiti from the time it was ceded to France by Spain in 1697 until the return of Jean-Bertrand Aristide in 1994 has been of the social conflicts of a population stratified by pigmentation. The whites hated and feared the mulattoes. The mulattoes and the whites hated and feared the enslaved blacks. In 1791, the slaves, seething under this combined oppression, began a rebellion almost continuous until independence was granted in 1804. The success of the uprising was largely due to the charismatic leader Toussaint L'Ouverture, who rose from the black, despairing masses. He was born a slave and named Francois Domenique but later adopted the *nom de guerre* L'Ouverture ("The Opening"). He was self-educated and became a brilliant tactician; he made life a perilous hell for the whites and mulattoes.

The revolt in Haiti threatened the grand designs Napoleon Bonaparte had on the American continent. It was his intention to consolidate French holdings from New Orleans through the Mississippi valley; Haiti was vital as a base to stage his administrative and military forces for the mainland. The rebellion had to be put down, and in 1802 he sent 60,000 troops, the flower of the French army, under the command of General Charles Victor Emmanuel Le Clerc to Haiti. But in 1802 yellow fever, always smoldering in Haiti, raged in epidemic form. From all accounts, the strain of virus was highly virulent causing great mortality to the unseeded non-immunes; a physician there called it the *coup de barre* ("blow with a stick"). Within the year, 23,000 troops died of yellow fever. A dispirited Napoleon came to view America not as a source of wealth and power but as a worthless, pestilential sinkhole. In 1803 he instructed his minister, Charles-Maurice de Talleyrand, to take whatever he could for his vast, bad real estate. Talleyrand sold it to the United States for $15 million.

While epidemiologic ignorance of yellow fever had the serendipitous effect of bringing Louisiana into the U.S. fold, that ignorance was bringing New Orleans and other affected regions of the South to their economic knees. The erroneous belief in contagion—the direct jump of yellow fever from person to person—led the populace and authorities of the day to the conclusion that exclusion was the way to prevention. On learning of yellow fever in a town or city, all commerce and travel was denied. It was city against city; Mobile would isolate Memphis, and Galveston would isolate New Orleans. Roadblocks, manned by vigilantes were thrown up to turn away travelers and goods. During this early- and middle-nineteenth-century yellow fever time in the United States, the railroads were also beginning their phenomenal expansion. In 1840 there were about 2,800 miles of track; twenty years later this network had grown to some 31,000 miles. Merchants were quick to adopt this new rapid and relatively inexpensive way of moving goods and produce. However, when yellow fever struck, the railroads were viewed not as a channel of commerce but as a channel of contagion. Railroad vigilantes were formed that prevented trains from stopping in their yet-uninfected town. If their effort to keep the train moving failed, the vigilantes resorted to the simple and very effective expediency of tearing up the tracks.

Maritime commerce fared even worse, choked off by quarantine of the port cities. Sailing ships bringing slaves and trade from the known, notorious foci of infection—West Africa, South America, Havana—were considered as the sources of epidemics. Ships coming from the places of entrenched or epidemic yellow fever were met by the authorities and were forbidden entrance if there were any sick crew members or passengers aboard. Quarantined, (the term literally meant 40 days but the prohibition varied from place to place and time to time). The quarantined ship had to fly a yellow flag from its mast. This flag known to sailors as the yellow jack soon came to be synonymous with yellow fever, and by the mid-1800s yellow jack became the common sobriquet for yellow fever.

The quarantines imposed against ship, railroad and road began to unravel the economic integrity of the young country's southern states. New Orleans, a major entry and transhipment port for South American and Caribbean agricultural products, was especially hard hit by the disease itself and its economic consequences. The city became the U.S. pariah, the nucleus from which yellow fever was believed to spread. Ships began to bypass New Orleans and sail directly to New York and other northern ports where the merchants had enough political clout to have the quarantine less strictly enforced.[20] Not only was trade disrupted, but money for capital development of industry in New Orleans and the South was being denied by the cash-rich, hard-hearted northern bankers who would not make loans to a sick city or state any more than they would to an ailing individual. A sanitarian of the period, Elisha Harris, trying to deal with the direct and indirect economic effects of yellow fever, commented that, "The health of the Southern States cannot suffer a visitation of yellow fever without a loss of millions, at once in the Banking and Commercial interests that are centered in the City of New York." A present-day historian, Margaret Humphreys, has estimated that the reduced capital investment cost New Orleans $10.5 million each year from 1846 to 1851, a time when yellow fever was highly prevalent and virulent in the city.

The dream of a new South made prosperous by industry and international commerce was thwarted by yellow fever. Immigration bringing new people and new ideas into the South virtually ceased. King Cotton was dethroned and exiled; somehow the belief arose that cotton was the carrier of yellow fever. Some southern cities refused to accept the cotton crop for processing or transhipment. Because of this policy the economy of the stricken Memphis, Tennessee, became so parlous that the city declared bankruptcy and repudiated all debts—a singular, desperate civic act caused by a disease. A municipality in fiscal straits these days would simply issue new bonds. The citizens of Memphis considered stronger medicine—that their city should be put to the torch. Fortunately, cooler heads prevailed, but even so, depopulation contributed to and compounded the economic troubles.

20. Also, for reasons that are not clear, yellow fever outbreaks ceased to occur in the northern United States after about 1850.

In the American Decameron people fled from yellow fever. In 1841, Florida had lost so much of its population that Congress seriously considered delaying its admission as a state. White owners abandoned their plantations, leaving the black slaves as caretakers. Yellow fever gave the slaves of the U.S. South their chance to revolt, but no Toussaint L'Ouverture emerged to lead an insurrection for freedom.

Clearly, the quarantine laws were not having their intended salutary effect. Merchants were losing money. Worse still, those merchants were forced to pay the costs of imposing the quarantine. Yellow fever was forcing the entrepreneurs to become active players in the health field. The business community lobbied the authorities to relax quarantine enforcement. They argued that the quarantine laws were an unconstitutional limitation of interstate commerce (finally, in 1886 the Supreme Court took up the matter and rejected the argument). The New Orleans press did its civic duty by not reporting the number of yellow fever cases. However, the most effective ploy of business was to mount a campaign to deflect attention from quarantine to contagion. Contagion was better for business; it was manageable. The business community's propaganda was that the real cause of yellow fever was a noxious agent in filth and fomites. Prevention of epidemics was only a matter of cleaning up the city. And if the elected officials were unable to impose civic hygiene, the business community would do so. The merchants of New Orleans formed their own Auxiliary Sanitary Association whose candid slogan was Public Health is Public Wealth. Large amounts of private money went to cleanup projects—collecting garbage, sweeping streets, and so on—but yellow fever returned, again and again, unabated.

The social mischief of the contagion theory is that if filth causes disease, yellow fever in this case, then some group must be making the filth and endangering the clean, prosperous, (Protestant) God-fearing, U.S.-born citizens. The contagion theory demonizes and you don't have to look far to spot the demons—immigrants, African Americans.

The Irish were the first to be demonized. They had begun their immigration in the 1770s, but the great wave of over a million people was in the 1840s when the mass starvation caused by the potato blight drove them to the United States. Of all immigrants coming to our shores at that time the Irish were most discriminated against—reviled for their poverty, feared for their Catholicism. It was "common knowledge" that the "dirty Irish" were the source of yellow fever; in 1855, at the apogee of an outbreak in Norfolk, Virginia, a mob burned the Irish ghetto.

New Orleans, vortex of yellow fever in the South, contained too few Irish to make them the accused. But it did have lots of African Americans. The African Americans were particularly suspect because they didn't sicken and die with the same great frequency as the whites; their partial resistance led the logic of that day to the conclusion that they were the source, the carriers of the yellow fever contagion. A growing community of Italians in New Orleans also didn't meet the patrician standards of cleanliness or racial intolerance. They too came under suspicion. A New Orleans physician, C. M. Brady, voiced the public opinion of biracial epidemiological causation when he stated that the African Americans "associate with Italians on certain terms of social equality."

Those first two hundred years of yellow fever in the United States are an object lesson, of timeless validity, of how a swiftly devastating disease striking in epidemic proportion can bring a nation to economic crisis; destroy civic, social, and personal ethics; and evoke the most brutish bigotry. The next two hundred years also provide an object lesson of timeless validity. It is a lesson of how science and brave people can rescue a country and a world in microbial peril. That rescue began with the research of a French chemist who began to ponder on why his wine had turned so disgustingly undrinkable.

Chapter 6

The Germ's Warfare Discovered: 1650 to 1865

THE HUMAN SENSES are pathetically feeble. Only with prostheses of technology have we been able to peel the unseen universe to its cores—the earth and cosmos in its atomic particle structure; the molecular transactions of the living cell; and, most important to human health, the invisible world of microbes.

The man who gave us first sight of the teeming minuscule cosmos of life was the Dutch draper of Delft, Anton Leeuwenhoek. Why this mid-seventeenth-century merchant-burgher became a hobbyist obsessed by magnification we do not know. The new art of lens making gave rise to the first and fundamental understanding of the laws of light. With the crude tools of the day, Leeuwenhoek acquired the skill to grind the small pieces of glass into lenses for his microscopes.

I recommend Delft in midwinter when the grayness and chill give the city, much of it unchanged since Leeuwenhoek's day, a charming monochromatic patina. To warm you, an Indonesian restaurant serves the hottest food I've ever had, and in the movie houses coffee or liquor can be brought to your seat. At city's edge there is the ultramodern European Primate Center where the Leeuwenhoekean pursuit of "animalcules" continues. Here Dr. Alan Thomas and his group are giving chase in experimental chimpanzees to that elusive quarry, the malaria vaccine. Perhaps it was those nasty, rainy winters of Delft when even the canals reflected the somberness of the city that drove the meticulous Leeuwenhoek to the solitary and painstaking pursuit of lens making. Good microscopes he did make. Even today's optical experts have judged them, considering the technology of the time, to be remarkably fine instruments.

One day, probably in 1675, Leeuwenhoek placed a drop of pond water under his microscope and saw exploding into view, a field swarming with swimming, blobbing, dancing creatures—"animalcules" he was to call them. It was like looking up into a sky that had been solidly black at one moment and seeing it alive with stars at the very next moment. In all the years of his 91-year-long life that excitement of exploring the microscopic world never left Leeuwenhoek. At 88 he described himself as being in the "autumn of life." He examined everything—shrimp heads, pepper infusions, the insides of insects, frog feces, his own feces—and reported his findings in hundreds of letters to the Royal Society of London. There is a sense of breathless wonder when he writes, "In the month of June I met with some frogs whose excrement was full of innumerable company of living creatures of different sorts and sizes. . . . The whole excrement was so full of living things that it seemed all to move." But although Leeuwenhoek opened that invisible world to view, neither he nor anyone else for the next 200 years realized that some of those animalcules could make you sick, some, deathly sick. The essential connection that was to change biomedical thought and practice so profoundly had to await the genius of Louis Pasteur.

In 1994 I had the privilege of speaking ("Malaria and Motherhood") at Paris's Pasteur Institute. It was the celebratory Pasteur Year at the institutes he founded and continued to bear his name. Even in death, he never left his beloved mother institute; his remains lie in the basement. My friend, colleague, and guide, Dr. Geneviève Milon took me to this holy of holies, Pasteur's crypt. While walking from her laboratory, we shared some "trade" gossip—who was doing what in the malaria research business; did the newly published results of the malaria vaccine trial in Africa represent a great advance or an overstatement of "massaged" data? Geneviève confided that the British science establishment's nose was out of joint. The Pasteur Institute had built a new library with money from the sale of the Duchess of Windsor's jewels she had bequeathed to the Pasteur. The Brits thought that the money should have been theirs; the jewels had been bought with the pounds sterling of a member of their Royal Family, albeit a somewhat tarnished member. I said the money should have been the United States's; after all the Duchess had been a U.S. citizen. Our casual banter came to an end when we entered the quiet splendor of the Pasteur crypt where we came to pay our respects to the father of our science. The crypt's walls are adorned with mosaics of stolid goddesses, depicting the three basic virtues, Faith, Hope, and Charity. Another, very Pasteurian virtue goddess had been added to the conventional trio—the goddess Science.

From the sanctum of the crypt we walked upstairs to the museum of Pasteur's life and work. There is a reconstruction of his apartment, solidly middle class. There was the surprise of finding Pasteur the artist—the portraits in pastel that he did as a teenager. They were remarkably fine and professional, and I thought how lucky we were that he turned to science rather than to art (curiously, he never put brush or pastel to paper again after his teen years). Then there was the large room that held the relics, in clockwise temporal progression, of the extraordinary diversity of his research.

The room of relics reflected Pasteur's rather pantheistic devotion at Science's altar. It was not his style to grub away in a narrow corner of research; the progression from the chemical nature of tartrate to the rabies vaccine may seem disconnected but was, in fact, a logical progression of research.

Louis Pasteur (1822–1895) was first a chemist and a teacher. Professor of chemistry at Paris's Sorbonne from 1867 to 1869, he forged a trail that led from chemistry to microbial pathogenesis. The diversity of his insights made up his genius.

Those who would ask for your support in the war against this or that disease would have you believe that science is very focused; all the eggs need be in one grandly funded basket to win the war. This is of course partly true; directed research is necessary, but it is also true that one doesn't know from whence the essential problem solver cometh. Our remarkable advances in immunology spring from learning the function of a small bit of tissue in a chicken's bowel. Similarly, the unlocking of yellow fever and all other microbial diseases stemmed from Pasteur's discovery of what made a wine go bad. Pasteur's progression from a sick wine to a rabid dog is a lesson in the interconnectedness of science.

It began as pure chemistry, a discovery that the crystals of tartaric acid exist as two forms, one bending light one way, the other bending light the other way. Pasteur now asks where tartaric acid and its relatives—malic, lactic, and butyric acids—come from. He finds that they are natural end products of sugars such as in grape juice or apple juice, but what drives the reaction from sugar to the acid and other end products (such as alcohol) is unknown. Here, the long-dead Leeuwnhoek provides the magnification for Pasteur's vision. Under the microscope Pasteur sees swarms of animalcules in the grape juice that is about to become wine. These animalcules are identified as yeast cells, and when Pasteur kills the yeast with heat (a process now called Pasteurization of milk and other liquid foods), the grape's progression to wine, that subtle mélange of alcohol and organic acids, abruptly comes to a halt. This leads Pasteur to the conclusion that, "The true ferments were organic beings."

The national fermentation, the pride of France, was the wine made by all those special French yeasts. The truly bad wine of 1864—a national calamity—was virtually undrinkable. Armed with his newly gained knowledge of the fermentation process, Pasteur, a true patriot and lover of good wine (the Pasteur Institute makes its own wine from grapes grown on the institute grounds; it is sort of Beaujolais-like and very pleasant even when drunk young) turns his attention to that impossible vintage. He discovers that a "rogue" yeast is making that abnormal fermentation. The wonderful Pasteur intelligence seizes on the concept of sickness, be it of wine or women, as an abnormal fermentation mediated by "rogue" pathogenic microorganisms. That conclusion leads to a remarkable series of researches and discoveries. The beer goes bad and another aberrant yeast is identified. The silkworms are dying; the French silk industry is crashing. A protozoan pathogen (a microsporidian) rather than a yeast is proven to be the culprit in this case.

For Pasteur, it is a short step from the sick silkworm to the sick animal. Under the microscope bacteria are seen causing a form of cholera-like diarrhea in chickens, of anthrax in sheep, and of boils in humans. The Germans enter the germ game and, led by the formidable genius of Robert Koch (1843–1910), discover still other causative pathogens such as the mycobacterium of tuberculosis. Ways to culture bacteria in special media are invented as are new ways to stain and taxonomically place the species of bacteria. Bacteriology as a discipline is well established by 1880, and for the next 30 years there is an extraordinary enthusiasm for bacterial villainy. It is an era of bacteriomania; virtually all human ills are thought to be of bacterial causation. All one has to do is be diligent enough in the search and the offending microbe will appear under the microscope. But a few illnesses demurred; they had all the marks of the microbe but no visible microbe. That research trail, first traveled by Pasteur, led from rabies to yellow fever and all other viral diseases.

In the Old Curiosity Shop that is the Pasteur Museum, there is a treasured relic of medical history. It doesn't look like much—a ropey bit of dried up tissue. It is, in fact, the desiccated spinal cord of a rabid rabbit. Rabies needs no further elaboration other than the reminder that the untreated bite of a mad dog (or other rabid animal; it spares no mammal) will cause the most horrible of deaths. The French are intensely devoted to their dogs. Rabies has always been endemic in France (a major reservation the British had of building the Chunnel was the thought of a rabid French dog or fox tearing down the tunnel to contaminate the rabies-free British Isles). Rabies was a challenge to Pasteur.

In 1882 Pasteur was inducing rabies in rabbits by inoculating into their brains an emulsion made from the spinal cord of mad dogs. The rabbits developed typical rabies and died. He then passaged the spinal cord emulsion from a dying rabbit into a clean rabbit and noticed that with successive passages the disease became hotter; the rabbits died more and more rapidly after getting the intracranial inoculum. Obviously the agent of rabies was being passed in the inoculum from rabbit to rabbit and becoming more virulent at each passage.

All evidence and instinct pointed to a microbe of rabies,

but the microscope and culture—using the methods of that day—were blank. Later, a ceramic filter was devised to sieve out even the smallest bacterium. Emulsions of rabid rabbit and dog brains were passed through, but the clear "sterile" filtrate was still capable of inducing rabies when inoculated into an experimental rabbit or dog. Either rabies was a non-microbial disease or the causative organism was so almost unimaginably minute that it was beyond the magnification of even the most powerful of light microscopes. Pasteur believed in the too-small-to-see explanation, and his next experimental maneuver was based on that assumption.

The ever-practical Pasteur never took his eye off the true objective; the discovery of a new microbial pathogen was all very satisfying, but to be of value, it must lead to cure or prevention. For rabies even the starting point, the identified microbial cause, was absent and certainly there was no cure by nostrum then or now. The missing microbial link didn't deter Pasteur. He could follow Jenner and his own experiences in dealing with rabies; he would vaccinate (the term "immunization" came later).

An English country doctor, Edward Jenner, made the astute observation that milkmaids never got smallpox. Cows got poxlike pustules on their udders; milkmaids got pustules on their hands, which cleared up without treatment. Following self-cures, they then had lifetime resistances to both cowpox *and* smallpox. Smallpox was a constant, uncontrollable threat not only in England but throughout the human-habitated world. Surely, reasoned Jenner, what was good for milkmaids was good for everybody; and to prove his point he began to inoculate cowpox from milkmaids into indigent children from the workhouse. Then to prove his hypothesis, he had his nonphysician nephew inoculate the real thing, material from smallpox pustules, into these children. Fortunately, Jenner was right and the youngsters were protected. The year was 1796. Jenner didn't know from viruses or antibodies; but he now knew that a course of a mild disease, cowpox, would protect against contracting the related severe, sometimes fatal disease, smallpox. That was vaccination.

Pasteur had a somewhat related result with chickens. He had isolated the causative bacterium, now known as *Pasteurella antiseptica,* of fowl cholera and began to grow them in a nutrient broth he had formulated. It was cheaper than chickens, and when he did want to induce infection, it was a simple matter of taking a dose from the culture and giving it to the hapless fowl. However, the bacteria demanded a careful husbandry and the culture would crash when the rapidly dividing organisms exhausted the medium's nutriments. Relatively rapid subpassage to fresh culture medium was necessary to keep the strain going in the "test tube" *(in vitro).* On one occasion Pasteur delayed subculturing just a little bit too long and when he tried to induce fowl cholera with these effete bacteria nothing happened, the chickens remained alive and healthy. Not a man to let a chicken go to waste, Pasteur later gave these chickens a dose of virulent bacteria from a fresh culture. Again nothing happened; the chickens had been, somehow, protected by having first been dosed with the weakened bacteria. This serendipitous "failed" experiment provided the crucial clue that a modified pathogen could induce resistance to a disease. In this case it was experimental chickens and the fowl cholera bacteria. It was 1887. Almost one hundred years earlier Jenner did much the same thing with cowpox and smallpox in his experimental children.

The rabies organism couldn't be grown *in vitro.* It couldn't even be *seen.* But Pasteur was convinced that it was there, lurking in the experimentally infected rabbit's nervous system, and that à la chicken cholera, he could induce protection against rabies with a weakened organism. After a long series of experiments, beginning in 1882, Pasteur inactivated morsels of spinal cord from a rabid rabbit by suspending them for an optimal length of time in dry air. Then in an another long series of experiments he proved that dogs could be made refractory to rabies with a series of daily inoculations, over a week or two, with the inactivated spinal cords. In three years he had solidly "immunized" (it was immunization but Pasteur had no concept of the mechanism of the immune reaction) 50 dogs. It was 1885 and it was show time.

The boy, nine years old, was named Joseph Meister and he was terribly frightened and in great pain. Two days before, on July 4, 1885, he had been attacked by a rabid dog belonging to the town grocer, Théodore Vone. There were numerous deep bites on the boy's hands, legs, and thighs. The local physician had cauterized the wounds with phenolic acid, but there were so many bites that despite this painful procedure, the boy would surely become rabid and die of the hydrophobia.

Now, on July 6, the desperate, panic-stricken mother had brought her child from their home in Alsace to Louis Pasteur's laboratory and the one person she believed could work the miracle to save her son. On that very day Pasteur attended a meeting of the Paris Académie des Sciences where he told two distinguished physicians, Prof. Graucher and Dr. Vulpian of the boy's plight. He also told them of his successful experiments with dogs in preventing rabies and suggested that they boldly proceed with the first human trial and treat the boy. Pasteur later wrote, "The death of this child appearing inevitable, I decided, not without lively and great anxiety, as may well be believed, to try on Joseph Meister the method I had found constantly successful with dogs." Graucher and Vulpian agreed and hurried to Pasteur's laboratory and the waiting child. That evening, at 8 o'clock, they injected, under the skin of the boy's abdomen, the first dose of a preparation made from the dried spinal cord of a rabbit that had died of rabies 15 days before.

Thirteen of these inoculations were given over a period of ten days. On the tenth day the ethical, devout Pasteur did something that we in the post-Nuremburg age would consider criminally outrageous. Like Jenner before him who challenged his vaccinated children with the virulent smallpox virus, Pasteur challenged Joseph Meister with "the most virulent virus of rabies" that was taken from an experimental dog injected with material of rabbits "which produces rabies after seven days incubation."[21] Pasteur excuses the act with the lame reasoning, "The final inoculation with the very virulent virus has the further advantage, it puts a period to the apprehensions which arise to the outcome of the bites." Well, maybe; but I think Pasteur's curiosity as a scientist clouded his ethical judgement. Meister was another, albeit final, experiment and he wanted incontrovertible proof of its conclusion. At any rate, young Meister lived.

21. Pasteur uses the term "virus" in a general way as an unseen virulent agent. From the time of discovering the filterable virus agents until the early 1940s, viruses were identified, classified, and crystallized but never visualized "in the flesh." With the invention of the electron microscope which had enormously greater magnification powers than the conventional light microscope, the virus was finally observed as a "real" object. There is some difficulty in tracing the first sighting, but I believe it was of the vaccinia virus in 1942 by R. H. Green and his colleagues.

Chapter 7

Yellow Jack and the Cuban Crisis: 1885 to 1900

BY 1885 the base of microbiology had been cast. Miasma was out; microbes were in. It now became possible for medical scientists to reconsider yellow fever. The disease had abated in the North but continued in epidemic spurts in the South, Cuba, and the tropical Americas. Quarantines and chemical disinfections were having no beneficial effect; the bacteriologists were the best new hope. Unfortunately, the overenthusiasm for bacteria and the technical inability to carry out the rigorous experimentation needed for proof of causation led to numerous microbial red herrings of yellow fever.

In 1885, the year of Pasteur's triumph over rabies, a Dr. Domingos Freire of Rio de Janeiro announced that he had "bagged" the microbe of yellow fever. It was a bacterium he named *Cryptococcus xanthogenicus.* He cultured it, "aged" it, attenuated it in the manner of Pasteur, and inoculated this "vaccine" into several thousand people. It had no protective effect against yellow fever, but at least it was relatively harmless; only one person died from an adverse reaction. It eventually turned out that *Cryptococcus xanthogenicus* was really the inelegant *Staphylococcus aureus,* a common cause of boils. The year 1885 also witnessed the discovery of another bogus yellow fever organism. Dr. Manuel Carmona y Vale of

Mexico announced that the etiological agent was a spore he named *Peronospora lutea* in human urine. He dried the urine in which he found the "spores" to make a vaccine and proceeded to inoculate several thousand people with decomposing urine. "The results were what would be expected," as a reviewer, Dr. E. K. Sprague commented in 1914.

And so it went. In 1887, a Dr. Gibier of Havana finds a bacillus in the corpses dead of yellow fever; Dr. Carlos Finlay also of Cuba finds a (false) coccus bacterium, but he redeems his mistake by becoming the first true prophet of how the disease is transmitted. In 1896 a Dr. Sanarelli of Brazil and Uruguay isolates the ultimate yellow fever red herring which he names *Bacillus icteroides.* He gets a horse and over a period of 18 months repeatedly injects the animal with cultures of this organism (it is uncertain what the bacterium really was, possibly it was *Corynebacterium diphtheriae,* the cause of diphtheria). Then he takes the serum from the animal and injects 15 to 65 cubic centimeters of it into each of 22 patients with yellow fever. Again, as expected, there is no effect.

A U.S. Army surgeon, George Sternberg, the military microbiologist who began the crucial assault on yellow fever, commented, in 1890, on the useless and dangerous etiological assertions: "Among the microorganisms encountered there is not one which by its constant presence and special pathogenic power can be shown indisputably to be the infectious agent in the disease."

The Cuban rebels–guerrillas–freedom fighters operated from a U.S. coastal town from where they planned to invade and free their country from its tyrannical government. If this sounds familiar, it's a case of mistaken historical identity; this was not 1961 and the disaster of the Bay of Pigs but 1897 and the disaster of yellow fever in Ocean Springs, Mississippi.

By the late nineteenth century Spain's empire was coming apart, but it still managed to maintain a steadfast hold on Cuba. It was an oppressive colonial administration forcefully putting down any opposition. Throughout the 1800s the Cubans rebelled, fought, and schemed to free themselves from Spain's rule. The Americans sympathized with their cause; from 1776 onward we have always been supportive of democratic revolutionary causes elsewhere. In the case of Cuba the altruism was leavened with base politics and commercial ambitions. Before the Civil War the southerners envisaged Cuba's becoming another U.S. state—a slave state. After the Civil War the northern industrialists envisaged Cuba as both a market for their manufactured goods and a source of raw material. Cuba is still potentially a country rich in agricultural products and mineral reserves.

And so, the Cuban freedom fighters, remnants from a major insurrection led by the poet José Marti, were welcomed when they came to Ocean Springs, a resort town on the Gulf of Mexico. That welcome turned to outraged rejection when yellow fever broke out shortly after the Cuban's arrival and spread to other towns along the Gulf coast. Cuba had long been considered by the Americans as a health threat, an epicenter from which yellow fever spread. Here was ample confirmation of that belief. The Cubans had come carrying the yellow fever germ (or whatever it was) and spread it to the unsuspecting citizens of Ocean Springs (in this opinion those citizens may have been right). An undercurrent of opinion had long held that the United States should take over Cuba for medical reasons, a yellow fever cleansing. The sinking of the *Maine* was just the ticket to do so.

It was a splendid little war—sweet, short, and profitable. On February 18, 1898, the American warship *Maine*, anchored in Havana harbor, blew up with a loss of 250 lives.

It is still not certain whether a mine or misadventure caused the explosion, but goaded by the yellow-dog newspaper publishers William Hearst and Joseph Pulitzer, the event inflamed the bellicosity of the U.S. public and its Congress. On April 19 Congress told President William McKinley to go ahead, take Cuba by force. Before McKinley could act, on April 24 a furious Spain, in desperate machismo, preempted him by declaring war on the United States. Shortly thereafter 17,000 U.S. Army troops assaulted Santiago, Cuba. This siege almost failed when yellow fever struck the invaders, but within a month Santiago fell. Another 16,000 troops invaded Puerto Rico where Lieutenant Colonel Theodore Roosevelt, the second in command of a peculiar volunteer cavalry group, called the Rough Riders and made up of football players, playboys, and cowboys, charged up Kettle Hill. A continent and ocean away Admiral John Dewey took his squadron into Manila Bay.

By December 10, 1898, the parties were in Paris signing a peace treaty. Spain lost everything; Cuba became an independent protectorate under the United States. The United States got Puerto Rico outright. As always, the United States bought real estate at the bargain prices of a distress sale; for $20 million it bought Guam and (an unwilling) Philippines from Spain. The United States had joined Europe as a colonial power and, like the European imperial countries, assumed the problems and burdens of the subject land, people, and parasites. The European powers had to deal with malaria, sleeping sickness, schistosomiasis, meningitis, cholera, and other myriad diseases of the tropics. The United States in the tropical Atlantic now had to deal with yellow fever.

The catalyst in the investigations that finally solved the yellow fever problem was George Miller Sternberg (1838–

1915), a small town boy from Hartwick Seminary, New York, who took his M.D. from Columbia University's College of Physicians and Surgeons. Sternberg was to become a distinguished U.S. pioneer in bacteriology and then, in 1893, surgeon general of the U.S. Army. He and Pasteur had both discovered the diplococcus bacterium of pneumonia about the same time, and he had a residual sense of competitiveness with the French microbiologists. When an Italian, a bacteriologist at the Paris Pasteur Institute, announced from Uruguay that he had discovered the cause of yellow fever, Sternberg viewed the announcement with a skeptical eye. The bacteriologist was Sanarelli and his alleged pathogen was *Bacillus icteroides.* In December 1898, almost immediately after the conclusion of the Spanish-American War, Surgeon General Sternberg dispatched Dr. Aristides Agramonte to Havana to refute or confirm Sanarelli's claim.

In Agramonte, Sternberg had the ideal man in Havana. He had been born in Cuba but raised in New York City after his revolutionary father was killed by the Spanish and the family had to flee. Like Sternberg, Agramonte took his medical degree at Columbia University, and like Sternberg he became a bacteriologist. When Sternberg's call came, Agramonte was an assistant bacteriologist in the New York City Health Department. He was commissioned as acting assistant surgeon and assigned to Military Hospital No. 1 in Havana where he set up his laboratory. Two years of hard labor in the laboratory, wards, and morgue produced nothing. Tissues and bodily fluids from the dead and dying of yellow fever yielded bacteria, but none could be specifically associated as the cause of the disease. Sanarelli was wrong. Sternberg was right. It was time to move on to new directions and a greater effort.

Headquarters of the Army,
Adjutant General's Office,
Washington, May 24, 1900
34. By direction of the Secretary of War, a board of medical officers is appointed to meet at Camp Columbia, Quemados, Cuba for the purpose of pursuing scientific investigations with reference to the infectious diseases prevalent on the Island of Cuba. Detail for the board:
Major Walter Reed, Surgeon, U.S. Army;
Acting Assistant Surgeon James Carroll, U.S. Army;
Acting Assistant Surgeon Aristides Agramonte, U.S. Army;
Acting Assistant Surgeon Jesse W. Lazear, U.S. Army.
The board will act under general instructions to be communicated to Major Reed by the Surgeon General of the Army.
By command of Major General Miles
H. C. Corbin
Adjutant General

Agramonte was to stay on as head of the laboratory in Havana, but leadership was to pass to a regular army officer, Major Walter Reed. Reed was an officer and a gentleman and a very bright bacteriologist. Although born in Virginia and not one of the Columbia University Physician and Surgeons coteric of bacteriologists, Reed was yet another New York City adoptee, having gone to Bellevue Medical College for his M.D. He joined the army and with Sternberg became a founder of U.S. military microbiology.

Ever since conflict resolution began to be carried out by the engagement of masses of professional soldiers, military leaders have recognized that diseases could be more dangerous than the enemy. By the turn of the century, in the new post-Pasteurean age, the military of most of the countries with big war machines had created facilities and/or operational groups to conduct pertinent medical research, particularly in microbiology. Some of the best scientific minds came into the military and some of the best research on infectious diseases came, and still comes, from military research (although today the obverse side of that research is biological warfare). In the United States there was the Army Medical Museum (later to evolve into the Armed Forces Institute of Pathology), of which Reed was curator when appointed to head the Yellow Fever Commission. As head of the commission Reed made crucial decisions of tragic consequences, but the trust and faith of his colleagues in the enterprise never wavered. Agramonte may have been the scion of a Cuban aristocrat, but he was also a New Yorker and had the acute perception of a streetwise kid. I trust his evaluation of Reed when he wrote of him, "Reed was a man of charming personality, honest and above board. Every one who knew him loved him and confided in him. A polished gentleman and scientist of the highest order, he was perfectly fitted for the work before him."

James Carroll was in some respects the most remarkable of the quartet. Born in Woolich, England, at the age of 15 he emigrated to Canada. When he came of age, he crossed the border to join the United States Army. His was not of the officer-class background; he entered as a private and was assigned to the Medical Corps (the Medical Service Corps in which we of World War II served). He became interested in medicine and determined to become a physician. It is uncertain how he did it, but evidently while still serving in the army, this man, nearing middle age, attended the University of Maryland Medical School. At the age of 37 he became an M.D. and went on to study bacteriology. Carroll was a quiet, intense man, "industrious and of a retiring disposition" as Agramonte, the group's unofficial chronicler, described him. He was also an exceptionally courageous man.

Jesse William Lazear, born in Baltimore, was a gentleman, a member of the Columbia club, a classmate of Agramonte's at that medical school. Agramonte thought the world of him: "A thorough university man, he was the type of old southern gentleman, kind, affectionate, dignified, with a high sense of honor, a staunch friend and a faithful soldier." Lazear, a career army man who had been given special training in malaria and mosquitoes, arrived in Cuba several months before the board was convened to head the Columbia Barracks Hospital Laboratory. Lazear too was an exceptionally courageous man. He was to be martyred by yellow fever.

On June 25, 1900, the four met on the veranda of the Columbia Barrack's Hospital officers' quarters and reviewed Sternberg's charge to them. They were instructed to investigate yellow fever, malaria, leprosy, and (in their spare time) fevers of unknown origin. It was like someone from on high giving an order to solve the mysteries of God, man, and the universe in two years. Fortunately—infectious disease hunters can have a very different sense of what passes for good fortune—a severe outbreak of yellow fever occurred just as the board came to Cuba. Several U.S. soldiers died. Others, recorded as dead of malaria by an army doctor acutely befuddled by his opium addiction, had, as Agramonte later showed, been killed by yellow fever. Circumstances thus gave priority to yellow fever. Reed's problem lay in deciding the priority of direction the board's research should take. He knew he was dealing with an infectious agent, but he couldn't cope with the invisible and, at that time, the unculturable. As an infectious disease specialist, he also knew that the real name of the game is "interruption of transmission." You don't necessarily have to know *what* is being transmitted as long as you know *how* it is being transmitted to deal with an infectious disease. For yellow fever the board decided to put to the test the discredited mosquito theory of that Cuban crank, Dr.

Carlos Juan Finlay. In doing so, they also had to put the fomite theory to rest.

He could have been named Charles John, his father was Scotch and his mother French, but he was born in Cuba so it was Carlos Juan Finlay (1833–1915). Although schooled in France, he came to the United States to study medicine (M.D., Jefferson Medical College, 1855) and then returned to practice in Cuba. Sometime during the 1870s Finlay came to the conclusion that yellow fever was transmitted from person to person by the bite of a mosquito. He wasn't sure what was being transmitted—a microbial agent or a poison, the mosquito bite being kin to the bite from a venomous snake. The notion wasn't all that eccentric because the last decades of the nineteenth century saw the dawn of the Age of Insects, that is, insects of medical importance. In 1876 Patrick Manson, an English physician employed by the Chinese Custom's Service in Amoy, showed that the filarial worm parasite causing elephantiasis was carried by the mosquito. In 1893 the Americans Theobold Smith and F. L. Kilbourne published their paper on the transmission of Texas fever, a disease of cattle caused by the Babesia, a protozoan cousin of the malaria parasite, through the tick. These two discoveries established by experimental proof the principle that some pathogens underwent a mandatory developmental cycle in an insect or other blood-feeding arthropod (a tick is not an insect but an arachnid, like the spider). The phenomenon was confirmed on the fourth of July, 1898, when Ronald Ross in Calcutta nailed the mosquito to the malaria cross.

Finlay, who formally announced his belief in the yellow fever–mosquito connection in 1881, was never able to win the needed experimental proof. He also somehow got the idea that mosquitoes not only transmit the "agent" but that in the mosquito the "agent" was so transformed as to be immunizing. His guinea pigs were Catholic divines. God knows how he persuaded them, but over a period of 7 years, from 1883 to 1890, Finlay collected 65 Jesuits and Carmelites who came to Cuba as missionaries. Finlay fed mosquitoes (type unknown but probably some were *Aedes aegypti*) on yellow fever patients and then allowed those mosquitoes to feed on the missionaries. He did this with 33 fathers and kept 32 other fathers as "unfed" controls. Of course, this didn't induce any immunization, but luckier still, it didn't produce any yellow fever in the experimental fathers. They were probably saved by Finlay's (and everybody else's) ignorance of the facts of life regarding the yellow fever virus. Finlay may have fed the mosquitoes on the patients at the wrong stage of their infections and/or not allowed enough time for the cycle in the mosquito to "mature" to infectiousness before feeding on the experimental subject. Finlay never got these critical times right, but he never abandoned the mosquito as the theoretical purveyor of yellow fever and, if anything, over the years became ever more obsessed by his unproven hypothesis.

Reed and his group became sympathetic to Finlay's point of view after witnessing a very practical demonstration of yellow fever epidemiology. The shambles of misdiagnosis and mismanagement of the yellow fever cases in U.S. troops at Pinar del Rio was a concern to the chief surgeon headquartered in Havana, and he ordered the ever-efficient Agramonte to the scene. Agramonte quickly isolated the cases in tents in the surrounding woods and had them attended by soldiers who had survived yellow fever and were now immune. On July 21, 1900, Reed came to Pinar del Mar and, in Agramonte's words, "I well remember how, as we stood in the men's sleeping quarters, surrounded by a hundred beds, from several of which fatal cases had been removed, we were struck by the fact that the later occupants had not developed the disease. In connection with this, and particularly interesting, was the case of a soldier who had been confined to the guard house since June 6; he showed the first symptoms of yellow fever on the twelfth and died on the eighteenth; none of the other eight prisoners in the same cell caught the infection, though one of them continued to sleep in the same bunk previously occupied by his dead comrade." This paradox made Reed and Agramonte to consider "that some insect might be concerned in spreading the disease," and "it was decided that although discredited by the repeated failure of the ardent Dr. Finlay to demonstrate it, the matter should be taken up by the board and thoroughly sifted." The board met on August 1 and formally decided to pursue the mosquito. The only one among them with any entomological training was Lazear, and he was put in charge of the investigations. However, we should pause here to consider the state of medical entomology in 1900 so that we may appreciate the formidable task that lay before Lazear and his colleagues.

The unsung heroes of biology are the taxonomists who tell the experimentalist what organism they are studying and what its relationship is to all other organisms. Today, the powerful tools of genetics and immunology can be applied in the sorting and taxonomic assignment but the foundation of "what to call it" still rests on the morphological description. This takes an orderly mind, particularly when it comes to the taxonomy and identification of insects. Keying in an identity depends on anatomical minutiae—how the insect's hairs are placed and grouped, the formation of the mouth parts, the sex parts, the bewildering pattern of wing venation. And for insects with a complicated life cycle, such as the mosquito, this whole identity process has to be done for the egg, the four larval stages, the pupa, and the adult; each stage has its own specific characteristics. The late, great medical entomologist, Prof. Patrick Buxton of the London School of Hygiene and Tropical Medicine (a colleague of mine with strong Anglican convictions once told me he went to bed each night with the Bible and Buxton's *Louse* on his bedside table), attempted to introduce me into the mysteries of mosquito identification. It was hopeless; I just could not follow all those wing veins.

Mosquito taxonomy was still rather a mess when Lazear began his work in 1900, although he was in a better position than the pre-Linnaean naturalists and their free-for-all nomenclature. In 1758, the Swedish naturalist-physician Carolus Linnaeus, who had laid down the rules (by which we still abide) of binomial nomenclature, genus and species—*Homo* (genus) *sapiens* (species)—applied these rules in the Tenth Edition of his *Systema Naturae* to the naming of a mosquito specimen from Egypt. He called it *Culex aegypti;* 175 years later with the blessing of the highest court of taxonomy, the International Committee of Zoological Nomenclature, it received its official title, *Aedes aegypti.* During that 175 years it had been called by at least 24 other names. Every time a specimen came from a different part of the world, it was given a different name, and not for many years did it dawn on the entomologists that they were dealing with a single species that had a cosmopolitan distribution.

The board moved to Havana where Lazear began to raise laboratory-bred mosquitoes. He put them into individual tubes with netting at one end so that the mosquito could be applied to a blood donor's arm. Through July and early August, Lazear fed his mosquitoes on yellow fever patients and two or three days later offered the mosquitoes a feed on a volunteer; the volunteer, as often as not, was a casual passerby, most likely a private in the army. Those soldiers thought it was all a big joke, but Reed, Lazear, Agramonte, and Carroll knew differently; they knew it was a deadly game in which the "winner" would get yellow fever. It was Finlay's work all over again, and as with Finlay all the trials, all the feedings, failed; not a volunteer came down with the disease. And the watching Finlay, so anxious to be vindicated, was becoming irate.

Then on August 27, Lazear and Carroll were in the laboratory, Lazear complaining that one of his charges, fed on a yellow fever patient 12 days ago, now refused all blandishments to feed again. Lazear demonstrated the reluctant mosquito by putting the tube to his arm without whetting the mosquito's appetite. Then the phlegmatic Carroll put the tube to his arm; the mosquito came down and inserted her hypodermic needle proboscis, and her belly began to swell with blood.

Two days later Carroll had a fever. Like so many physicians who temporize in diagnosing their own illnesses, Carroll first thought that nothing was really wrong, he had merely become chilled after a swim in the ocean. Then as the fever persisted and intensified, he took the diagnostic refuge for tropical fevers—it was malaria. That next day Agramonte and Lazear found Carroll in the laboratory hunched over the microscope, vainly searching his stained blood film for the malaria parasite. His two friends were aghast. This was not malaria; they saw the bloodshot eyes and suffused face that were the signs of yellow fever. Carroll was put into the hospital bed to await the uncontrollable outcome of his disease—death or recovery—while Lazear and Agramonte were left to ponder how the infection had been acquired. It was some hours before the stunning fact hit them; in Carroll's perilous illness there was a dark victory—it was the mosquito! Maybe! The experiment had to be repeated. The mosquito that had been fed on Carroll was still alive and it was again hungry.

Fifteen minutes after Lazear and Agramonte came to the conclusion that the mosquito had transmitted the infection to Carroll, their next human guinea pig, Private William H. Dean, Troop B, Seventh Cavalry, walked past the laboratory. Dean, salutes the two officers.

Lazear: "Good morning."

Dean: "You still fooling with mosquitoes, Doctor?"

Lazear: "Yes, will you take a bite?"

Dean: "Sure, I ain't scared of em."

Five days later Private Dean joined Assistant Surgeon Carroll as a patient in the yellow fever ward. Fortune smiled on both men; they would recover. Lazear, Agramonte, Reed (still in Washington), and Finlay were delighted; their mosquito transmission theory had been proved without incurring a fatality. But the virus was soon to impose the death penalty on one of them.

On September 16, Lazear was back in the laboratory feeding his mosquitoes. As he held a tube to a yellow fever patient's stomach, a "wild" mosquito flew in and came to rest on his hand. Lazear didn't want to move and unsettle the mosquito in the tube so he watched while "his" mosquito bit, took blood, and flew off. The wild mosquito was infected with the yellow fever virus; five days later Lazear became ill; six days later, on the twenty-fifth of September 1900, he died.

The three remaining members of the board were numbed, grief stricken by the death of Lazear. Sometimes, in collaborative research an almost-loving bond is formed (and sometimes, a mutual hatred of similar intensity). And so it was with the yellow fever board; they spoke of Lazear as a martyr. His sacrifice should have been adequate proof of the mosquito as the transmitter of yellow fever. Reed, Agramonte, and Carroll were convinced, but would the scientific community accept their evidence? There was only the one well-controlled experiment, that of Private Dean; the infections acquired by Carroll and Lazear were highly suggestive of mosquito transmission, but that evidence was, nevertheless, circumstantial. Was Dean enough? Was it necessary to put other lives in peril for the sake of convincing the doubters? When is enough enough in experiments on living, sentient creatures, be they mice, monkeys, or humans?

The most impressive "sheepskin" I've ever received was not for an academic degree (in my case, the higher the degree, the smaller the document) but a vivisection license from the British Home Office that allowed me to experiment on everything but "horses, asses, and mules." It was elegant—large, embossed, beribboned, and signed by a peer of the realm. It was issued only after the Home Office inspector was satisfied that I was able to handle experimental animals in a humane manner and that the animals were maintained under optimal conditions. Only when I had that license did my professor allow me to begin my dissertation research.

The necessity for keeping on the good side of the animals and the inspector was impressed on me by an incident that occurred shortly after I began graduate study and got my vivisection license. The inspector didn't like the way a reader (an associate professor without teaching duties) in the department was keeping his monkeys. The reader, a South African Boer with a short-fuse temper, responded to the criticism by punching the inspector on the nose. His license was yanked, and his experiments came to an end for almost a year before he groveled enough to have it reissued. During that time, he would come to the lab in the dead of night, conduct secret experiments on hedgehogs, and leave their wretched corpses on the desks of the graduate students.

In the succeeding years since the reader punched out the inspector, the scrutiny on the care and use of experimental animals has become even more searching. Now, protocols must be submitted to institutional or academic committees on animal experimentation for their approval even before a grant can be submitted to the funding agencies. The funding agencies have their own guidelines for animal care that must be met, and the federal overseers must be satisfied that the animal facilities meet their standards. Believe me, getting experimental studies underway can be very complicated. And expensive. One of my colleagues, whose grant had just been cut, suggested that it would be cheaper to keep her mice at the local Hilton hotel and give them room service than to pay the per diem charges of the university's Laboratory Animal Service.

As you can imagine, the restrictions and safeguards on the conduct of experiments on humans are even more rigorous than for experimental animals. There are several added layers of committees to satisfy, and in the United States the FDA and, often, the National Institutes of Health are the final arbiters of what can be done to whom. Written consent forms are required, but even with that permission, the lawyers are watching for legal and professional transgressions.[22] Although there may be mistakes and misdeeds, we live in a remarkably moral and ethical age of biomedical research. The abominations revealed at Nuremburg made all scientists who used humans as experimental subjects reconsider their procedures, and from this came the protective standards we have followed for the past half-century. Having abided by

22. For those who do field studies on tropical diseases, the written consent form can get a bit tricky when dealing with people who can't write, let alone read. In one instance when we were drawing blood for malaria and filariasis investigations on the Hagahai, a newly contacted hunter-gatherer group in Papua New Guinea, they not only couldn't read or write but spoke no known language. Regulations were preserved by two interpreters who went through three languages to and from Hagahai to the Melanesian pidgin which we could understand. Thumbprints were made on the form and our various committees were, I hope, satisfied.

those standards all our professional lives, how then should we now judge the medical researchers of another time? They used people in their experiments as if they were but guinea pigs; and yet those scientists were not monsters; they were honorable men, dedicated scientists. It is a dilemma that confronts me with confusion and sadness each time I come to write of human experimentation of an earlier time. I have never been able to sort it out, nor can I do so now as we again take up the yellow fever story and Reed's uncompromising pursuit of absolute proof of mosquito transmission.

Reed did the honors; in October he read a paper at the American Public Health Association's meeting in Cincinnati describing the board's findings. It was a perfunctory exercise; no acceptance of the theory was demanded, as Reed and the other remaining board members had already decided on carrying out additional transmission experiments. In this they were aided and encouraged by the military governor of Cuba, Brigadier Leonard Wood. Wood was a military oddity, a physician (Harvard M.D.) who became a "real" soldier after joining Roosevelt's Rough Riders. He ended his career as the U.S. Army's chief of staff. Wood recognized, as no monosoldier could, the importance of yellow fever as a problem affecting the health and efficiency of his troops. So, in a way, he became a committee of one on human experimentation in approving Reed's request to recruit new volunteer subjects.

After the death of Lazear, Agramonte became the designated entomologist and assumed the task of breeding and feeding the mosquitoes in his laboratory colony. In November, a secluded location was selected as their yellow fever research encampment. Seven tents were erected, a guard posted, and the in-memoriam name of Camp Lazear bestowed. To that camp, as the first subject, came the most heroic, self-sacrificing John R. Kissinger, a private in the Hospital Corps. He asked for no reward of money or rank and offered himself, as Agramonte notes, "in the interest of humanity and the cause of science." On December 5, 1900, Kissinger took the bite of mosquitoes that had fed, about a week before, on a yellow fever patient. Three days later he had yellow fever. The course of the disease was relatively mild and he recovered.

Shortly afterward another volunteer came forward. Unlike all the other subjects who were military, J. J. Moran was a civilian working for the army. He too asked for no reward. Moran entered a screened room in which there were infected mosquitoes. He lay down on a bed and allowed the mosquitoes to attack him. On Christmas Day, 1900, Moran showed the first symptoms of yellow fever. He too had the good fortune to survive.

Now there were Dean, Kissinger, and Moran, all experimentally infected under well-controlled conditions by the bite of *Aedes aegypti* mosquitoes. In addition, there were Lazear and Carroll whose infections supported the mosquito transmission hypothesis, albeit in a more circumstantial fashion. Even considering the then nascent state of statistics as applied to the pronouncement of experimental proof in biomedical research, one would think that three good examples and two semigood examples would constitute proof enough especially when inducing a disease in humans that had a 30 percent mortality rate and no known therapeutic cure. The five were not enough for Reed, and in the early months of 1901 he and his board were searching for more volunteers.

Soldiers have their own survival wisdom and strategies in war and in peace. A paramount and ancient principle that even I learned in my first week of basic training is that you never, never volunteer for anything. In the subculture of the other ranks you also learned to heed the scuttlebutt. The scuttlebutt among the soldiers in 1901 Cuba was that letting those crazy doctors feed mosquitoes on you was no longer a joke. That's the way you got yellow jack and maybe died. So the military pool of potential subjects was, for the most part, no longer available.

Now Reed and his colleagues took a page from Finlay's book and sought the Spanish immigrants who were still coming into Cuba. These were not priestly people, who presumably also learned of the true consequences of the mosquito's bite, but poor, naive peasants coming to find a new life in Hispaniola. These volunteers were bought volunteers. Here is how the candid Agramonte described the collection of the human "material" the board wanted for its experiments. "We thought best to secure lately landed Spaniards, to whom the probable outcome of the test might be explained and their consent obtained for a monetary consideration. Our method was as follows; as soon as a load of immigrants arrived, I would go to Tiscorina, the Immigration Station across the Bay of Havana, and hire eight or ten men as day laborers, to work in the camp." These hirelings were given easy chores and fed well for a few weeks until the guinea-pig proposition was put to them. Agramonte continues, "Naturally, they all felt more or less that they were not at all averse to allow themselves to be bitten by mosquitoes: they were paid one hundred dollars for this, and another equal sum if, as a result of the biting experiment they developed yellow fever. Needless to say, no reference was made to any possible funeral expenses."

When the fourth experimentally infected Spaniard was carried from Camp Lazear to the hospital, the other volunteers-in-waiting panicked. Reed made the wry, cold remark that those fleeing Spaniards "lost all interest in the progress of science and incontinentally severed their connection with

Camp Lazear." That put an end to the human experiments; it was finally proof enough for Reed and his board. The essentials of mosquito transmission were now known: (1) transmission from human to human was through the bite of the *Aedes aegypti (Stegomyia)* mosquito; (2) the mosquito became infected only when it fed on the patient during the early stages of the infection, the first three days after the onset of symptoms being the most important period; and (3) there was an incubation phase in the mosquito, taking about a week (depending on temperature), before the mosquito was capable of transmitting the infection.

That's all there was to it; one puny mosquito to be eradicated and with it yellow fever. To the U.S. military here was an ideal enemy: it didn't like to campaign too far from its mess quarters; a finicky eater, it couldn't forage the country it occupied; and it demanded very special conditions for its domestic life. Translated into entomological sense, *Aedes aegypti* was an urban, domestic mosquito, preferring to feed on humans, not usually flying and disseminating very far, and breeding in small collections of water, such as ditches, water cisterns, and empty cans. These behaviors made it vulnerable; have your sanitation squads clear or deny breeding water, screen over containers essential for water holding, and, when possible, screen the houses. This ideology of sanitation, environmental hygiene, is neither high tech nor fashionable in our present quick-fix world. Not many of our youths are willing to take up the career of sanitary engineer or sanitarian ("shit house wallahs" the British colonials ungenerously called them). Nevertheless, given an amenable mosquito species, it will work, but it is hard work and needs a determined advocate.

The U.S. Army had its determined sanitarian in William Crawford Gorgas (1854–1920). Gorgas, another southerner, from Mobile, went north for his medical education to Bellevue Hospital Medical College (there is no Bellevue Hospital Medical College as such today; it eventually became incorporated into New York University School of Medicine). A year after graduation he joined the U.S. Army Medical Corps and was posted to Fort Brown, Texas, where he came down with yellow fever. Gorgas recovered, and the episode sensitized him to the peril of yellow fever in a way that no textbook account or even patient care could. And it made him an ideal soldier to send to Cuba; now solidly immune against yellow fever, he could dare tread where the immunologically naive would be in harm's way. He was also one tough cookie who didn't mind throwing his weight around to get what he wanted. Today, Gorgas is considered to be something of a U.S. hero, but by today's standards of social morality, he would also be considered to have been a bigoted martinet. He once said that in their later years his wife and presumably himself also could be quite happy as owners of "a hundred bales of cotton and forty or fifty ebony faces."

Cuba in 1901 was, for all practical political purposes, a military dictatorship ruled by the U.S. Army. With a grand latitude of authority Gorgas set about imposing antimosquito decrees. Any householder found to have the *Aedes* mosquito breeding on his property would be fined $10. His "sanitary police" patrolled the city to enforce cleanup measures. At the same time Gorgas had his own workers vigorously engaged; they oiled the surface of breeding waters, suffocating the aquatic but air-breathing larvae; drained ditches; and screened over the water cisterns so cherished by the female *Aedes aegypti* as places to lay their eggs. Some miles from Havana, in Santiago, a new method of mosquito control was introduced. "Mosquito fish" (Gambusiae/mollies), voracious feeders on mosquito larvae, were put into collections of waters such as ditches and small ponds. It was effective then, as it is today—a low-technology, nonchemical way of dealing with insecticide-resistant mosquitoes. A mere year after Gorgas began this great work, yellow fever was banished from Havana and was rapidly vanishing from the rest of Cuba.

Gorgas's selective mosquito control was to be adopted by other countries throughout the twentieth century. Seventy-five years later Singapore, a democratic dictatorship, attempted to rid that city-state of its mosquitoes. Gorgasian sanitary laws were passed, with suitable stiff fines to encourage the citizenry. However, booming, bustling 1975 Singapore wasn't as ecologically amenable as 1901 Havana; the rainwater-filled excavations created during construction of the many new buildings made ideal, intractable breeding sites for the aedine and culicine mosquitoes. Nor did 1975 Sri Lanka, a socialist dictatorship, do much better. There too antimosquito ordinances were passed in trying to deal with the urban, *Culex fatigans*–transmitted filariasis (elephantiasis). But the Sri Lankan citizens refused to take any notice of the regulations. Mosquito cleanup was not for them; that's what socialist governments were for.

The twentieth century thus opened with a medical flourish and not a little bit of smug complacency. Humans had proved themselves superior to the *Aedes aegypti* mosquito. Yellow fever had been mastered and was about to be expunged in a rational way. Intense antimosquito campaigns began in the affected cities of the South, often with money from a public subscription.

The federal and state governments, though, were still not meaningfully engaged in protecting the health of the public. When yellow fever struck New Orleans in 1905—an outbreak that killed 452 people—the city fathers supplicated the Great White Father in Washington, President Theodore Roosevelt, for assistance from the newly created Public Health Service. The Public Health Service responded to the emergency by declaring that it was broke and could do nothing. The citizens of New Orleans then raised $250,000 to pay for mosquito cleanup and to screen in the yellow fever wards. With the onset of the cool weather of December the mosquito population further diminished and there were no more cases of yellow fever—ever! It was the last time yellow fever would ever occur in the United States of America. In the cities of the Caribbean and Central and South America similar operations were taking place and yellow fever was on the decline. There was the vision that it would be the first infectious disease to be eradicated. But yellow fever was a slumbering, bestial virus lurking in unthought of places and animals. It was to return and kill again.

There was, however, a respite from yellow fever, at least in the "civilized" world. In their complacency the Americans felt little need to develop strong expertise in diseases of the tropics. For the United States in the first two decades of the twentieth century tropical medicine was mostly military medicine, as it had been since Congress voted $300 to buy quinine for George Washington's malarious army.[23] The incubator of biomedical scientists, medical school teaching-research departments, had not been established. In 1900 medical schools were mostly vocational, turning out basic

23. That $300 may have been a crucial factor in the winning of U.S. independence. During the Revolutionary War both sides were affected by malaria but for the Americans it was a familiar American disease; they had the experience and wisdom to treat their volunteer army whereas the British largely ignored the malaria in their professional-mercenary troops. The British troops at the conclusive battle at Yorktown were observed to be in a pitifully malarious state (1780 and 1781 seemed to be highly malarious years with a severe epidemic in Philadelphia in 1780). Some medical historians believe malaria in General Charles Cornwallis's army assured General George Washington's victory.

physicians with basic skills. In 1910 a U.S. nonphysician-educator, Abraham Flexner wrote his wake-up report on the deficiencies of medical education in the United States and Canada. Medical schools began to change until they developed into their present form with basic medical science departments providing students with the knowledge of the scientific basis of medicine, training research doctoral students, and carrying out the bulk of U.S. medical research. While there are few tropical medicine departments as such, a great deal of contemporary research in the field has been performed by microbiology departments. But in 1900, except for pioneer universities like Johns Hopkins, departments doing that kind of research did not exist.

It was a different matter in imperial Europe where England, France, Germany, and Belgium had a vital interest in the warm girdle of the world. Trade in slaves and ivory had by then given way to trade in the rich commodities of the colonies—minerals, tea, coffee, spices, palm oil. Nigerian palm oil was particularly valuable. It made the British Lever brothers enormously wealthy, and their company was insinuated into virtually every cranny of Nigerian economic life. Nigeria was like a company town with everyone purchasing from, and in debt to, the company store. The French, like the British, had their colonial [illegible] throughout the tropics; they ruled in Africa, Asia, the Pacific, the Caribbean, and South America. The Germans ruled in East and West Africa as well as in a few islands in the Pacific. The Belgians had their Congo Free State—later the Belgian Congo, and now Zaire—owned as a commercial enterprise whose CEO was no less a personage than King Leopold who dreamed of an empire extending from the Congo to the Nile. Henry Morton Stanley played an important role in the forced recruitment of natives for Leopold's workforce. He commented that "the driving force of the Congo Free State was enormous voracity." Minor players of colonialism abounded too—Spain, Portugal, Italy, and Holland.

Those colonial powers faced a larger problem than merely maintaining a temporary military presence in their tropical possessions in contrast to the United States in Cuba and the Philippines (where it stayed longer than it expected and where it was not particularly welcome). The European concern was for the long-standing well-being not only of the occupying troops but also of their administrators, merchants, missionaries, and, for many colonies, colonists. Nor did the European colonial powers wish to rule over a population too sick and demoralized to work efficiently. This rude economic motivation mixed with the peculiar do-goodism of the era, a belief that colonialism was the best thing that ever happened to the "pagans." Joseph Chamberlain, the British secretary of state for the colonies said it was "constructive imperialism," and he sincerely meant it.

For all these reasons, the Europeans quickly realized that they required a medical service for their tropical colonies, and young medical school graduates were recruited for a life of foreign service. It quickly became apparent that these young doctors, fresh from European medical schools, didn't know very much about disease, sanitary needs, or health maintenance in the tropics. (Despairing teachers of the neglected study of tropical medicine in present-day European and U.S. medical schools might add, "So what else is new?") And then they realized that little was known about the practice of tropical medicine.

The closing decades of the nineteenth century and opening decade of the twentieth century witnessed the great discoveries—the mosquito transmission of malaria, yellow fever, and filariasis; the snail transmission of schistosomiasis; the trypanosome as the cause of African sleeping sickness and the tsetse fly as its vector. All this helped in disease prevention, as evidenced by the successful antimosquito campaigns against yellow fever. But if you were a patient in the tropics in 1900, only three drugs were available for treatment—quinine for malaria (still used and remarkably efficacious even now in dealing with multidrug-resistant malaria), mercury for yaws and syphilis (a really wicked, toxic therapy obviously no longer used), and ipecacuanha for amebic dysentery (a drug made from the root of a Brazilian plant *Psychotria ipecacuanha,* it had been used since 1658 but was quite toxic and of questionable value).

The Europeans addressed their colonial health problems by establishing great institutions of teaching and research where they trained their physicians and soon the physicians from throughout the world in tropical medicine. In 1899, the London School of Tropical Medicine opened its doors at Albert Dock, sharing the building with the Seaman's Hospital Society. Later it became the London School of Hygiene and Tropical Medicine and moved to Bloomsbury.

That same year, a few months earlier than the London School, the Liverpool School of Tropical Medicine began its long and distinguished institutional career. The Liverpool School was largely funded by the commercial interests of that city, a center of the profitable trade with West Africa. Sir Alfred Jones, chairman of the Elder Dempster Line, was a driving force in the founding of the school. The Elder Dempster Line was the major passenger-shipping line to the British colonies in West Africa. In 1951, I disembarked in Lagos from their ship the *Apapa* for my first steps on African soil. It was a two-week journey. The food was atrocious; I found only the mortadella baloney to be edible, and that turned a curious green by the time we docked in Lagos. The green mortadella made me question the passage in the West African pidgin bible describing the Flood, "Noah's ark be plenty fine, fine past Elder Dempster Line."

The continental colonial powers soon followed England's example and established their own teaching-research institutions of tropical medicine—Germany in Hamburg, Belgium in Antwerp, France in Paris, and Portugal in Lisbon. Even tiny, noncolonial Switzerland would have its Institute of Tropical Medicine. Moreover, these were not teaching-research schools isolated by geography and pure theory. They established satellite laboratories in the tropics where faculty could work and then, enriched by its experience, return to its home base.

Sometimes that experience was peculiar. The Liverpool School of Tropical Medicine had a field station, the Sir Alfred Jones Laboratory, in Sierra Leone. Sometime in the 1930s two young men who were to become distinguished professors of tropical medicine, Dr. R. M. "Dicky" Gordon and Dr. A. Blacklock, were carrying out research there. One evening after a hard day's work, they turned their attention to consuming a copious amount of pink gin. Blacklock decided he had to "go see Africa" and tottered to the outhouse to use the "thunderbox." He didn't take a lantern, and as he lowered his pants and was about to take the seat in the murky darkness, he experienced two quick, sharp stings on the buttocks—a snake! In panic he ran back to the house where Gordon found two neat puncture marks—the serpent's teeth. Through the fog of pink gin the pair tried to think of their next move and vaguely remembered that cutting and bleeding out the venomous blood was the way to go. Gordon, a sly, shy Milquetoast of a man. splashed some gin on the dignified Blacklock's behind, made the incision, and let the blood flow, and both had another postsurgical pink gin while awaiting any signs of the snake bite. After a time, with Blacklock seeming to be unaffected, they decided it would be prudent to identify the snake and kill it. Taking a lantern and shotgun they went to the outhouse. There on the seat of the thunderbox sat the chicken who had defended itself against the large, white mass descending on it with two successive sharp pecks.

It was only the United States that failed to develop a special center of expertise. Piecemeal, departments dealing with infectious tropical diseases were established in universities and research institutions such as Tulane, Louisiana State, Johns Hopkins, Harvard, the National Institutes of Health, the Walter Reed Army Institute of Research, and the Rockefeller Institute, but there never has been a center entirely dedicated to the study of the health problems of the people in the tropics. Perhaps one reason for this failure in centralization of effort is a critical difference between Europe and the United States. Europe was a conglomerate of northern temperate countries holding tropical dependencies whereas the United States was and is, in good part, a tropical country. Get a map of the world and look west from London; that latitude will take you to Hudson's Bay and the polar bears. Now look east from Miami; that latitude will take you to Africa and the elephants. England's essential tropical problems were in Africa and Asia; the United States's tropical problems were in Pennsylvania, Georgia, and the Carolinas. The U.S. doctors were treating their U.S. patients for malaria, dengue, yellow fever, and amebic dysentery. The English doctors were treating their English patients for gout and dropsy. The United States did eventually begin to deal effectively with the tropical diseases at home and abroad; the way was led by a Baptist billionaire and a blood-sucking worm.

Chapter 8

Great Works 1900 to 1925: Kid Rockefeller and the Battling Hookworm

WHEN MY WIFE and I lived in Antwerp in the fall of 1994, we grew accustomed to our daily exploratory walks, excursions sometimes interrupted by rain. On one such wet occasion we took refuge in the city museum. It was one of those funky little municipal museums often found in towns and cities throughout the western westernized world—places rarely visited, filled with local oddities. The Antwerp City Museum gave us an unexpected pleasure. It was having a special exhibit of regional culinary history—cookbooks and menus of notable banquets. We were delighted; we like to cook and we like to eat—although the inclusion of the *Molly Goldberg Cookbook* (in English) circa 1938 seemed a bit odd. One item especially intrigued and amused me, the menu of a banquet given on June 5, 1880, by the Societé Commerciale Industrielle et Maritime in honor of the French canal builder, Ferdinand de Lesseps. Fifteen courses, including Poulets de Mars a la Valoise—whatever that is, but with a name like that it must have been delicious. The menu was decorated with an engraving of the isthmus of Panama that showed a canal bisecting it.

De Lesseps had supervised the building of the Suez Canal, completed in 1869. Now he was going to dig a big ditch across the isthmus of Panama. With the canal the Europeans would regain commercial ascendancy in the Western Hemisphere. Those merchant-industrialists who gave that lavish dinner for de Lesseps should have saved their money; they put the canal before the horse. Yellow fever and malaria denied them the canal. Walter Reed's discovery of the mosquito transmission of yellow fever gave the canal—and all its enormous economic advantages—to the United States.

From 1881 to 1889 the French labored in Panama. They should have been aware of the health hazards of the region where they sent their malefactors to languish and die. The thousands of workers sent to Panama fared no better than the prisoners in the penal colonies of Devil's Island and Cayenne. In the eight years of construction, 22,000 French laborers died of disease, one-quarter of the workforce. Most fell to yellow fever although malaria and the dysenteries took their toll. Despite the fearful mortality, the French canal company extolled Panama as a tropical paradise in order to attract workers. Even Paul Gauguin was seduced by the advertisements and worked a miserable two months there with pick and shovel in 1887, until he was fired. But by then he had been captured by the heady life of the tropics. The rest is art history. By 1889 the French syndicate bankrolling and building the canal foundered near bankruptcy and was ready to throw in the spade. In 1903 the United States, for many years interested in owning a canal across either Panama or Nicaragua, acquired the rights to the Panama strip and its abandoned French ditch.

We moderns readily accept the new discoveries and whims of science—eat fish, don't eat fish; drink coffee, don't drink coffee. How, then, to understand the reluctance of even the highest U.S. government officials of 1902 to accept the fact that mosquitoes transmit yellow fever and malaria.

True, all agreed that without proper sanitation operations to protect its workers, the United States would fail, as had the French. But what form of sanitation? A large, influential body of opinion remained mired in the filth-contagion theory of yellow fever transmission. It believed that dirtiness brings the disease, not some insignificant midge. Fortunately for U.S. destiny the two foremost physicians of the day at the foremost U.S. medical school, the Canadian William Osler and William H. Welch at Johns Hopkins, had an interest in infectious disease and public health, and they espoused the mosquito as the target.[24] Together they promoted the Reed-Ross mosquito principle and advised President Theodore Roosevelt to appoint Gorgas as chief of sanitary operations for Panama where he would carry out a Havana-like antimosquito campaign.

Roosevelt really didn't think much of the mosquitoes or, for that matter, of the stern Gorgas (although he later pronounced him a "corker"), but he heeded the advice of Osler and Welch and in 1902 Gorgas was appointed officer in charge of health for the building of the Panama Canal. However, it wasn't until 1904 that Gorgas arrived in Panama and got down to serious work. He brought with him an outstanding team, the sanitary engineers Joseph LePrince and A. J. Orenstein, the director of hospitals Henry Rose Carter, and S. T. Darling as chief of laboratory and entomological services. He also brought with him a large sum of money; his $340,000 a year budget was, up to that time, an unprecedented allocation for a sanitation project.

Gorgas immediately began a massive assault on Panama City's *Aedes aegypti* population. He fumigated the city with

24. In 1918 Welch became the director of the first U.S. school of public health, established with Rockefeller money at Johns Hopkins. Osler left Hopkins to become Regius Professor of Medicine at Oxford and a "Sir."

sulfur and pyrethrum, oiled the breeding waters, and covered the drinking water cisterns. A year later, in April 1905, despite all these efforts the city was again stricken with an outbreak of yellow fever. The outraged governor, an unreconstructed contagionist, said that Gorgas should have been cleaning up the city not swatting harmless mosquitoes and demanded that Gorgas be replaced. Gorgas managed to hang on and was vindicated; the last case of that 1905 outbreak was the last case ever in Panama City.

Malaria proved a harder nut to crack. As the canal wormed its way across the isthmus it crossed swamps, streams, and jungle—an environment replete with ecological niches for the breeding of the malaria-transmitting anopheline mosquitoes. In contrast with the urban swat-team approach to the control of *Aedes aegypti,* the control of *Anopheles* (and malaria) required large-scale engineering works and the men and money to get them done. As many as 1,300 laborers worked in antimalaria operations across the isthmus. Swamps were drained and the bush cut back, but there still remained sufficient anophelines for the continuing transmission of malaria. Only when other, costly measures of screening the sleeping quarters and supplying bed nets were implemented did the incidence of malaria significantly decline. Sometimes brute force had to be applied against the mosquito. At one stage of the canal's construction the laborers slept in unscreened railroad box cars. The sickness rate from malaria was so high that work almost came to a halt. The ever-resourceful LePrince hired natives from the West Indies whose sole job it was to hand catch and kill, night and day, the mosquitoes inside the box cars. Malaria abated and the digging continued.

When the United States finally joined the Atlantic and Pacific oceans in 1914, malariologists and other tropical medicine experts throughout the world gave their resounding acclaim. Panama and Gorgas was the first proving model of everything they were advocating, that malaria, yellow fever, and probably other infectious diseases could be controlled by the species cleansing of antivector measures. Only the will was needed and they, the experts, would find the way. The ever-prickly, opinionated Ronald Ross castigated his government's failure to follow the Panama example: "Everyone knows how magnificently Gorgas maintained the health of the Canal Zone. If I had my way the same thing would have happened in every British colony."

In this noble objective the experts and their successors for the next 75 years lost touch with economic reality and biological facts. The economic reality demanded a lot of money for the vector control engineering works. Only when there was the prospect of profit would they be financed. The United States foresaw a quick fortune from the Panama Canal. Gorgas's budget was worth every cent—a neat, discreet health campaign that gave a high permanently positive bottom line. In contrast, the British, for example, would have no such economic advantage if they invested a proportionally large sum in malarious India or Sierra Leone. Money, however, could be made from a Panama-like antimalarial program in the British colony of Malaya.

The Malayan-British malariologist, Malcom Watson, made the pilgrimage to Panama. In the early twentieth century rubber was the wealth of Malaya. Vast tracts of jungle fell, replaced by rubber plantations. Great fortunes were made by the English owners and the Chinese immigrant middle traders. Neither the indigenous Malay nor the aboriginal Orang Asli had any interest in the laborious and painstaking work of the rubber tapper, but Malay's labor problem had been met by the importation of thousands of Tamils from south India. The descendent of one such laborer, the late Prof. Arthur Sandosham, M.D, Ph.D., a sweet man with an inexhaustible fund of quite terrible jokes, became the professor of medical parasitology at the University of Singapore where he carried out research on malaria. "Sandy" said his expertise in malaria was a tribute payment to his plantation worker forebearers who suffered so acutely from malaria. That was the problem that crossed social lines for the Tamil rubber tappers and the plantation owners. The more rubber planted, the more malarious the laborers became and the more the rubber production fell. There was a willingness to make the capital investment to control malaria and raise rubber production. They hoped that Gorgas's methods in Panama would show the way. It is worth looking at the Malaya experience since it began to reveal some uncomfortable realities: (1) even the best-funded strategy has to be based on good epidemiological-entomological intelligence; (2) malaria is largely a self-inflicted, human-made disease; and (3) ultimately money is the best antimalarial.

After Ross discovered that mosquitoes transmitted malaria and Grassi added the crucial observation that only Anopheles mosquitoes carried the human malaria parasites, it seemed that the conquest of malaria was but a campaign away. Kill the Anopheles by denying them their breeding water and malaria would disappear. Then it became apparent that there were over 300 species of Anopheles and only relatively few of those species were biologically and behaviorally capable of carrying human malaria parasites. The early simple concept of attack began to fall apart. Any campaign would require the identification of the one or more species of Anopheles responsible for malaria for that particular area. Only then could the species be tracked to her watery birthing pool and those waters drained or otherwise modified to make it unsuitable for mosquito larval life. Easier said than done.

In Malcolm Watson's day the mosquito detectives had a difficult time proving guilt. They could round up the suspects by hand catching or trapping. Then they could make them confess by pulling their heads off. The heads, salivary glands still attached, of these females (mosquitoes) would be taken to the crime laboratory where the salivary glands were examined under the microscope and guilt assumed when the infective form of the malaria parasites, the threadlike sporozoites, were seen. In most cases those pronounced guilty in this forensic way were indeed guilty, but sometimes condemning judgement was pronounced on the innocent. Watson too found it wasn't so elementary and was undone by an anopheline red herring.

When Watson's crew began collecting and examining the different species of anophelines in the rubber estates and surrounding jungle, they found one species, *Anopheles umbrosus*, to have a high sporozoite rate; that is, a relatively high percentage of mosquitoes caught and dissected would have the malaria parasite sporozoites in their salivary glands. Under the microscope all sporozoites look alike; they may be from other animal malarias, monkey malarias for example. Only quite recently have the exquisitely precise immunological and DNA-probing techniques made possible a species diagnosis of a sporozoite in the wild. But on the basis of the best circumstantial evidence Watson fingered *Anopheles umbrosus* as the culprit bringing malaria to Prof. Sandosham's grandparents and the other rubber tappers. So the crews went out and painstakingly drained the myriad small water-filled holes where *umbrosus* was breeding. Malaria continued unabated.

In Malayan folk stories the mouse deer plays a role like

Brer Rabbit or Wiley Coyote, a cunning but admirable con artist. The mouse deer is about as big as a big cat or terrier and can often be seen on the roads through the forests of Southeast Asia. It is also frequently seen in a curry or satay in country restaurants. About 50 years after Watson began his intense attack on *Anopheles umbrosus,* it was discovered that the mouse deer carried its own species of parasite and the mosquito that transmitted it from mouse deer to mouse deer was none other than *Anopheles umbrosus.* The mouse deer malaria had nothing to do with humans; *Anopheles umbrosus* had little or nothing to do with humans.[25]

25. The mouse deer has a prominent place in my recollections. In 1960 I arrived in Singapore to join the medical school faculty after nine years in Nigeria working solely on African trypanosomes (sleeping sickness). My old friend and London School of Tropical Medicine mate Arthur Sandosham was to relinquish the chair of parasitology to me and become vice-chancellor. I was met at the airport to be told that Sandy had given a speech condemning the dictatorial practices of Lee Kuan Yew, Singapore's authoritarian head of state. He then had to make a hurried departure to Malaya, and therefore I was to give Sandy's lecture tomorrow to the medical students on hookworms. Hookworms? I had been working on trypanosomes for nine years and had forgotten a lot of medical parasitology during that time. It was a busy week catching up.

During that frantic time I got a letter from my former professor at London Cyril Garnham, who was then writing his monumental book on the malarial parasites of man and animals. Garnham was an aesthete and he wrote, "Desowitz, find a classic Chinese artist and have him paint a scroll of a mouse deer for an illustration in my book." No matter what, old London professors are never to be denied and I dropped swatting my parasitology book and found the wonderful painter Chen Wen Hsi (now deceased as is Cyril Garnham).

Chen Wen Hsi was intrigued and agreed, but he was a city boy; he had never seen a mouse deer—he was really into gibbons which he painted beautifully, his models swinging freely through his house. We finally found a mouse deer in the sultan of Johore's private zoo. Wen Hsi painted it in a jungle bamboo grove setting, and to show his expertise, he painted it without a brush but with his fingers. And that's how the scroll of the mouse deer came to be in Garnham's book on the Haemosporidia. As I write this, I can look up above my desk and see Chen Wen Hsi's lovely finger painting of two cranes that he gave me.

The entomological penny dropped when it was found that another mosquito, *Anopheles maculatus,* was an important vector of the malaria plaguing the rubber estates. Now the female *maculatus* is genetically programmed to seek sunlit streams in which to lay her eggs. The shadowed jungle had relatively few suitable ecological niches for this otherwise efficient malaria vector. When the jungle was replaced with rubber estates, the sun was let in to shine on the thousands of streams and trickles coming off the Malayan hills. *Anopheles maculatus* proliferated, malaria intensified, and the more rubber that was planted the greater the number of malaria cases among the tappers and their families, and the plantation owners and their families.

Watson, with the financial backing of the plantation owners and the blessings of the colonial government, undertook the heroically difficult task of abolishing the streams by an extensive system of subsoil drains. In time, the *maculatus* numbers were reduced and malaria, although still present, abated. With some judiciously administered quinine, the tappers could be kept at work and the plantation owners could make a profit. In a way, it was a page from the Gorgas book of wealth through health. That page was not read by the Americans whose vast southland was economically blighted by malaria and other tropical diseases. It was through a diverse set of great philanthropies from one man, John D. Rockefeller, that the United States was brought to its medical senses and then dared to venture into helping people throughout the world in unburdening themselves from plagues and parasites. That great venture began with the sallow southerner and his hookworm.

John D. Rockefeller was an intensely devout Baptist, a man of the religious right who once told a reporter from the *Woman's Home Companion,* "God gave me my money."

When Rockefeller gave this interview in 1905, the income tax law of 1894 had been declared to be unconstitutional by the Supreme Court. Before The Tax a few men had amassed great wealth and many people had amassed great poverty. Lacking the will, philosophy, and money, the government did little to redress the plight of the industrial or agricultural impoverished worker; social justice was meted out through the philanthropy of the rich. It was philanthropy on a great scale, not the mere sending of a turkey in a basket to the ghetto at Christmastide. The rich thought it their moral duty to act as the social conscience. The billionaire Andrew Carnegie advised the billionaire John D. Rockefeller that, "The man who dies rich dies disgraced."

Rockefeller was reviled for his wealth and the sometimes ruthless way it was accumulated. There are those who believed and still believe that his great munificence was nothing more than public relations. This is an undeserved disparagement of Rockefeller's generosity. Throughout his working life, from his first job at 16 as an assistant bookkeeper until he became the owner of Standard Oil, he gave a proportion of his earnings to church and charity. In its beginning the big largesse was rather diffuse; Rockefeller was a soft touch for the Baptist churches and their missionaries in need and he gave as he was importuned. Nevertheless, he did have some notions, albeit still unstructured, as to where and for what he would like to see his money go. Although only a high school graduate himself, he believed in supporting higher education. Influenced by his wife Cettie, an ardent abolitionist, he became interested in the welfare of the African American.

It was a 39-year-old Baptist minister who put a focus to the Rockefeller philanthropic millions. It was a sinuous path of giving and organization that led from education to medical research and the heroic campaigns against tropical diseases.

The preacher, Frederick T. Gates, was secretary of the American Baptist Society when he met Rockefeller in 1888. The tall, handsome man—he looked somewhat like Billy Graham with a walrus moustache—was an eleemosynary genius. He also proved to have a good business head on his shoulders and became Rockefeller's trusted financial adviser. An early vision that Gates set before Rockefeller was the creation of a great religious university, a Baptist Oxford, from the then sorry ashes of the University of Chicago.

The University of Chicago was founded in 1859 but it had withered both in money and scholarship. By 1886 it fell into receivership, its mortgage foreclosed. Gates persuaded Rockefeller to save the university and, most importantly, its affiliated Baptist seminary to an amount that eventually totaled $35,000,000. And although it never became the Baptist Oxford, the University of Chicago did become today's great and innovative institution of learning. Higher education pried open Rockefeller's deep pockets; medicine and public health were soon to benefit too.

Gates again provided the impetus for philanthropy. In 1897 Gates, an omnivorous reader of catholic breadth, began to work his way through the 1,000-plus-page medical bible of the era, William Osler's *Principles and Practice of Medicine*. A medical fact struck the layman Gates—every ailment had a name and diagnosis, but almost none had a cure. Nature, the ability of the body to self-heal, was the best therapist; the physicians of 1897 had few effective therapies at their disposal. However, the close of the nineteenth century was a time when the magnificent three, the European scientists Pasteur, Koch, and Ehrlich, preached a hopeful message—the cures and preventions were there for the researching. Europe took heed; France, Germany, and England already had institutions for medical research. Gates believed that the

United States should do no less. Rockefeller welcomed the idea; after all he came from a medical family. His adored father, an itinerant self-styled physician, supported his family by offering to cure anyone of cancer for $25 "if they weren't too far gone."

In 1901 the Rockefeller Institute for Medical Research was incorporated with the promise of $20,000 a year for ten years. The institute had its first laboratory in a loft on Lexington Avenue. A year later, the enthusiastic Rockefeller, stimulated by Gates, gave $1,000,000 and bought the land the institute now occupies on the East River between 64 and 67 Streets. Abe Flexner's brother, the doctor Simon Flexner, professor of pathology at the University of Pennsylvania, became its first director. From its beginning the institute, later to become Rockefeller University, led medical research in a diversity of areas including investigations on the microbes and parasites causing "tropical" diseases. However, the institute's work was, as it continues to be, for the most part in rarefied, exciting research. Many of its findings devolved to practical application, but with rare exceptions, the Rockefeller scientists did not dirty their hands in the field labors of public health. Rockefeller philanthropy created another arm for that in response to the old human nemesis the hookworm.

Education was again at the headwaters of a health program. Both Rockefeller and Gates held that the impoverished African Americans of the South could enter the U.S. mainstream only if they were educated. Rockefeller and Gates believed that the chance should be given, especially so because most southern African Americans were Baptists; not white baptists, mind you, but Baptists nevertheless. In 1903, a General Education Board was incorporated with a little start-up money of $1,000,000 (by 1907, Rockefeller had given

$20,000,000 to the board). Whites were also to benefit, as the funds were to be given "without distinction of sex, race, or creed," but the greater portion went to African-American colleges, particularly Hampton College in Alabama.

Gates chaired the General Education Board and its secretary, soon to be its head, was another Baptist pastor, Dr. Wallace Buttrick. Gates and Buttrick made frequent trips to their educational "parishes," often traveling in Rockefeller's private railroad car. From the window of the interior opulence they saw a landscape of misery—pallid, listless white children, moon-faced children of both races supporting pot bellies matchstick-thin legs. Sometimes they were shocked to see children ravenously eating dirt! This was obviously not the human material on which to base their hopes for a better society through education. These two concerned, caring men understood that the health of the children must be improved if they were to avail themselves of the educational opportunities being offered. But what was making these dull-eyed people unhealthy, and what could be done for them? Not many physicians had an answer; a few doctors told them that the cause of the misery of the South was the hookworm and that the man who knew most about this parasite was working not in a medical institution but at the U.S. Bureau of Animal Industry; his name was Charles Wardell Stiles.

Stiles had studied parasitology at the University of Leipzig under the "renowned and immortal" helminthologist Rudolf Leukart. Europe had been mindful of the hookworm since the parasite was discovered in 1838 by Angelo Dubini in the cadaverous gut of an Italian peasant. The true importance of the worm struck Switzerland and its neighbors fifty years later with a virtual epidemic of hookworms in the laborers carving the St. Gothard tunnel through the Alps. The hookworm loads were so intense, the anemia so bad, that the tunnel had nearly to be abandoned.[26] Add trichinosis (the Europeans loved pork sausages), echinococcosis (they also loved dogs), and several other nasty worms. Europe had the parasites and it had the parasitologists to study them. A U.S. zoologist like Stiles who wanted to study parasitic worms went to Europe, where if you were very lucky, you might find a place in Leukart's laboratory. But when Stiles returned to the United States in 1891 with his brand new Ph.D., he had to find employment in veterinary parasitology. Americans had little regard, scientific or otherwise, for the parasitic worms of humans, but they were well aware of the great economic loss when worms slimmed their cattle and pigs or so emaciated the horses that they couldn't pull the plows. In Stile's time the center of U.S. helminthological research and expertise was at the U.S. Bureau of Animal Industry. From this Bureau, Stiles continued to study hookworms, and when he was called to meet Buttrick, he was the leading U.S. authority on the parasite.

Stiles and Buttrick met; Stiles talking throughout the night, explaining how hookworm oppressed the South and expounding on his vision of a hookworm-free, healthy, happy, prosperous, reinvigorated region. The next morning Buttrick went to New York City to tell Gates of his meeting with Stiles. Gates immediately enthused to the idea of an anti-hookworm crusade. In November of that year (1908) with

26. Hookworm anemia has long been a major disease of miners. In the late 1800s and early 1900s mines and tunnels had few sanitary facilities; miners defecated indiscriminately. In the warmth of the mine the hookworm eggs developed rapidly, and the men working at close quarters readily became infected, over and over. Hookworm was a serious disease of the California miner, especially at one digging where the boss was a Christian Scientist who believed in neither microbes or parasites. In France the shadow of the hookworm was still there in 1989 when cases were found in the coal miners of the Lorraine.

$1,000,000 of Rockefeller money, the Rockefeller Sanitary Commission was formed and began its operations from the command center in the heart of enemy territory, Nashville, Tennessee.

Nature is neutral; good and evil do not exist in its vocabulary. Humans make anthropomorphic judgements; we take our hurts personally. And so, when we look into the face of the hookworm, we see incarnate, menacing evil (figure 2). With its fangs and cutting plates it *looks* like the bloodsucker that it is.

When the barefoot boy steps on feces-contaminated soil, the lurking infective hookworm larva comes to the surface and penetrates the skin of the foot. It enters a small blood vessel and then makes its obligatory journey of life, like the salmon returning to the stream of its birth, through the heart to the pulmonary circulation of the lung where it ruptures the small blood vessels to enter the air sacs. The larva now continues its migratory passage up through the bronchioles

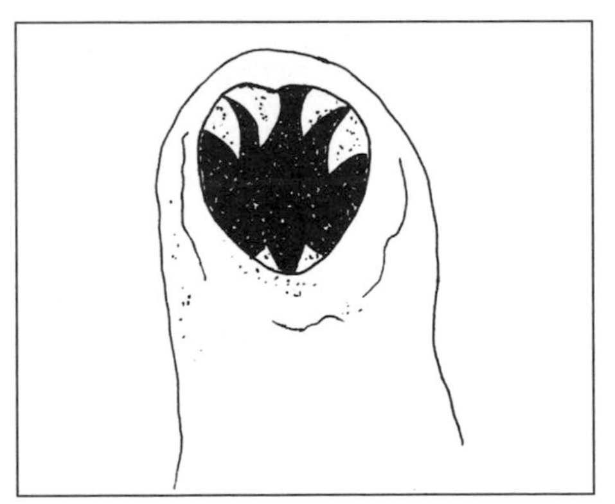

FIGURE 2
The face of the hookworm

and then up the windpipe (trachea). When the larva reaches the junction of the trachea and esophagus it . . . gulp . . . passes over into the digestive tract, descending unharmed and undigested through the stomach to the small intestine. Here it grows and mates; the female begins to lay 5,000 to 10,000 eggs every day for the five to ten years of her life. The worms clamp their "teeth" onto the gut wall, make a small lesion, and with their powerful pharyngeal pump begin their lifelong bloodsucking. All the worm really wants for its nutrition is the liquid part of the blood (the plasma); the red cells pass through unaffected but are lost all the same. Each worm takes from 0.03 to 0.15 milliliters of blood each day, and there may be a thousand or more worms in a heavily infected individual. Where there is no adequate makeup from food or supplement (such as ferrous sulfate tablets), the iron reserves that are needed to produce new, normal, oxygen-carrying red blood cells are depleted and there will be a profound iron-deficiency anemia.

Chronic hookworm anemia debilitates both the individual and the community. Children are stunted in growth and intellectual development. The adult's ability to work and provide food is diminished. The heart becomes enlarged in its exertions to pump enough iron(hemoglobin)-deficient blood to keep the body oxygenated. In heavily infected pregnant women, not given iron supplements, there may be insufficient oxygenated blood passing through the placenta to sustain fetal life. In the rural, hookworm-rife tropics this insidious worm causes more stillbirths than syphilis, eclampsia, or puerperal sepsis.

In 1910 the difficulties of conducting a public health campaign were not so different than now; as always, the irrational had to be overcome. But Wickliffe Rose, the epitome of the rational man, believed it reasonable that a sick person would

(1) want to know the cause of his or her illness; (2) readily take a curative medicine or other therapy; and (3) do everything necessary to prevent getting the sickness again or, if still uninfected, from getting it at all.

As to Rose's first belief, not even the physicians of the South were well informed about the hookworm, the disease it caused, its treatment, or its prevention, so you can imagine the abysmal ignorance of the rural African American and white uneducated poor. Knowledge, attitude, and belief surveys carried out by modern epidemiologists were not tools of the public health trade in 1910, so we don't know what southerners thought about their chronic hookworm disease if they thought of it at all. If you're never well, you may never know that you're sick.

Even some of the educated who did know denied and rejected the idea of a parasite-debilitated South. To them, southern pride and honor had been attacked by the Rockefeller carpetbaggers who came down from the North to tell them that they were a wormy, sorry lot and that they should clean up their act. Irving S. Cobb, the southern satirist, wrote deprecatingly about "Kid Rockefeller versus Battling Hookworm, winner take all." Irving S. Cobb was a fat man who obviously didn't have hookworms in his guts. And the Macon, Georgia, newspaper, the *Telegraph,* editorialized, "Where was this hookworm or lazy disease when it took five Yankee soldiers to whip one Southerner?" The Yankees won the war. Were the Rebs too hookworm anemic to fight at their best?

Rockefeller philanthropy ran on a belief in the power of education, and this became an important element in the Sanitary Commission's campaign. Getting the message out for public health education in 1910 was no easy matter. There was no radio, TV, ad agencies, or junk mail solicitation; there was only direct exhortation, town meetings, and the distribution of handbills. It was an incredibly massive effort of health education for its day or for any day for that matter. From 1910 to 1914 Wickliffe Rose's team organized 25 thousand meetings and exhibits, which were attended by 2 million people; they distributed over 3 million handbills. The people of the 11 southern states were taught how to make a proper privy; parents were exhorted to make their children wear shoes. Some of this massive public relations worked, but much of it fell on uneducated, suspicious deaf ears. And who could make children wear shoes in the rural south of 1910? Education needed the assistance of a judicious dose of salts, which brings us up to the second of Rose's beliefs.

When Rose began his campaign, he had only a one-drug salvo for his antiworm artillery, and that drug, thymol, had to be primed with a dose of salts. Thymol loosened the worms' grip on the intestinal wall but didn't expel them; for that a powerful purgative bowel mover was required. We have a description of this uncomfortable regimen from a doctor who was treating hookworm patients in Jamaica.

> On the afternoon before the treatment of hookworm infection the patient is given a preliminary purge of two capsules of Compound Jalap Powder between 3 and 4 P.M.
>
> On the day of the treatment, at 6 A.M., if the patient's bowels have moved, one-half of the prescribed dose of two capsules of thymol is given. At 8 A.M. the other half of the dose is given. At from 10 to 11 A.M., if the patient's bowels have not moved freely since the treatment was given, a dose of magnesium sulfate is administered.

This was obviously not the ideal regimen for mass therapy or for good patient compliance. One Jamaican patient was so moved by his chemotherapeutic experience that he wrote to the doctor:

> I write to complan erbout the hookWorm medsin. It is too stronge and has cause me a lot of trouble. From sins I take the medsin two treatment I feel stronge and well in boddie an min an like work more than before. The which is all very well.

But, Dr. Sir a grate change have come over me in other ways. From till I tak the medsin two treatment I was mild an of a sweet disposition an were known by sich throughout the country roun erbout. An very patient. Now sir all is chang and I feel quite civil to myself. So much so that on last Satday when one Jeptha Smith cuss me I box him down too hard and him now threaten to run law wid me.

Sir from your kindness if the aforesaid person run law wid me I respectful ask you to see His Honour for it were no other than the hookWorm medsin make me act in sich a fierce manner.

Expectin your kind help if need be an you indulgence I ever crave I

remain, as ever

Your patient,
Abijah Thomas

A few years later Abijah would have had a somewhat better "medsin." Thymol was replaced by carbon tetrachloride. That's the pungent liquid used by dry cleaners to despot and refresh your soiled garments. It was toxic and it did kill the hookworm. Unfortunately, it only irritated the Ascaris roundworms that usually were the codenizens of hookworms in the human intestinal zoo. These large, titillated worms tended to coil together into a mass, a bolus, so large as to obstruct passage of food through the intestine. In the early 1960s, when I was professor of parasitology at the University of Singapore Medical School, a time before that city-state had been "dewormed" by intensive sanitation laws and an excellent drug (mebendazole), the Ascaris bolus was the most common cause of acute abdominal emergency in children brought to the General Hospital.

The next drug to come along, hexylresorcinol, wasn't the sovereign hookworm remedy, but it was certainly better than cleaning fluid. Its discovery and implementation is worth the telling if only to illustrate the fact that often as not you just down know where the next drug will come from. The Departments of Pharmacology and Preventive Medicine of

Vanderbilt University were, in effect, the research arm of Rose and his Sanitary Commission. A Dr. Paul Lamson of the Department of Pharmacology had been screening a variety of compounds for anti-intestinal worm activity and found that members of the alkylresorcinol group showed promise. Lamson wanted to undertake human trials, but for this he needed more of the drug than his laboratory could synthesize. Then he learned that the pharmaceutical company Sharpe and Dohme (later to become the giant Merk, Sharpe and Dohme) of Philadelphia was not only making an alkylresorcinol compound, hexylresorcinol, but using it extensively in their mouthwash S.T. 37! At first, Sharpe and Dohme refused to supply Lamson with any of the chemical, possibly because they didn't want it to get around that their highly profitable mouthwash was a worm killer.[27] Later, Sharpe and Dohme relented and Lamson got his drug; it worked and Sharpe and Dohme added another product to their inventory, hexylresorcinol crystoids for intestinal worms.

The Sanitary Commission's campaign significantly reduced the burden of hookworm in the southern United States although complete eradication was quite impossible at that time.[28] The commission's work funded by Rockefeller's

27. To the 1950s or 1960s the academic biomedical scientists were the Brahmans and those productive researchers in pharmaceutical companies were the untouchables. Those in industry, as they were deprecatingly referred to, were even barred from membership in their professional society, the American Society for Pharmacology and Therapeutics. It may well be that this antagonism made the Sharpe and Dohme people unwilling to accede to Lamson's request.

28. The last resurvey that I could find in the literature was carried out in Georgia in 1972, sponsored, for some unknown reason, by the U.S. Army and confined to the examination of rural white children only. Even for this exclusive group the hookworm infection rates ranged, from locale to locale, from 1 to 12 percent. Mostly these were light infections, but some children still carried large numbers of worms and had hookworm disease.

philanthropy, despite the continuing malaria, malnutrition, and poverty, had made the U.S. South a better and healthier place. Equally important, however, was the profound effect that the campaign had in changing the way twentieth-century Americans thought about themselves and the world at large. In 1910, when the commission began operations, there remained the sense of Calvinist resignation, the religious-philosophical temperament of the founding immigrants, in the acceptance of disease and tragedy. It was God's will. Now that passivity was abandoned; this was a telling demonstration that disease could be fought not only by defensive measures of flight and quarantine but by energetic, active strategies using science's new and growing therapeutic armamentarium. It was a realization that human fate could be directed by human will.

In 1910, when the commission began its operations, the United States was still an insular, inward-looking country. We had fought wars beyond our boundaries, not really expansionist wars but rather wars waged to keep foreigners out of what we considered to be our turf—that turf being most of the Western Hemisphere. In 1910, the United States had no wish to be either a global power or a global benefactor. Rockefeller and his foundation were to be the wellspring that turned the United States, irrevocably, to its international responsibilities.

Gates and Rose were encouraged by the results of the hookworm campaign, but they understood that the U.S. South was a mere fragment of the parasite's global domain. Always the missionary visionaries, these two men were of the firm belief that it was their (and Rockefeller's) ethical duty "to extend to other countries and people the work of eradicating hookworm disease." In 1913 two organizational changes reflected the growing scope of the Rockefeller's international activities in public health. The Sanitary Commission was replaced by the encompassing Rockefeller Foundation which eventually came to support not only health projects but also a myriad of other activities in the arts, agriculture, education, and the soft and hard sciences as well as maintaining a sumptuous haven on Lake Como for working scholars. As part of the foundation there was an International Health Board, first to take care of the overseas hookworm projects but later responsible for all the foundation's international health activities.

The International Health Board undertook antihookworm projects in 52 countries on six continents. If there would ever be a Tropical Medicine Hall of Fame it would be filled with the portraits of the men chosen by Rose to manage the regional projects: Lambert in Oceania, Heiser in the Far East, Sawyer in Australia, and Strode in Brazil. But the Rockefeller money was never a handout; it was never a cynical ploy to gain an economic or political advantage without expecting anything to be done. In each country where the Rockefeller Foundation's Health Board operated, an agreement was exacted from the government that it would participate and furnish men, material, and other assistance as best it could. The partnership was important because it soon became globally evident, as it had in the U.S. South, that hookworm could not be completely eradicated or even controlled on a long-term basis by mass chemotherapy alone. Only sanitation, health education, and improved living conditions could ultimately solve the hookworm problem, and to this end the full cooperation of the people and their government was essential. The diplomacy, efficiency, purity of purpose, and expertise of the Rockefeller Foundation's

International Health Board should be the model and envy of all the present AID (Agency for International Development) or AID-oid international health operations.

The Rockefeller antihookworm demonstrations throughout the warm, wormy world were also a wake-up call for the labor-intensive megaagricultural, U.S.-dominated interests, epitomized by the United Fruit Company. For all practical purposes, the United Fruit Company was the government of Panama, Costa Rica, Colombia, and Nicaragua—banana republics in both the agricultural and political sense. The United Fruit Company profited enormously and was generally satisfied with the output of their banana-picking peasants. Now the Rockefeller Foundation's antihookworm campaign in the U.S. South had shown the company that it was, in effect, getting the labor from one-half to three-quarters of a peasant with the hookworm taking the profit from the other one-quarter to one-half. In the early 1920s the United Fruit Company began to emulate the Rockefeller antihookworm strategy. Its was a much easier task than that of the Sanitary Commission. The United Fruit Company had a dictatorial hold on the workforce and didn't have to go through any extensive rigmarole of health education. The peasants were told to take their medicine, defecate in the company-assigned sanitary conveniences, and wear company-supplied sandals.

In 1924 the United Fruit Company held a meeting in Panama to review the economic and health effects of the hookworm programs. It was a stellar assembly of the hookworm greats: Bailey K. Ashford of the Puerto Rico campaign, W. E. Washburn of the Jamaican campaign, Juan Iturbe of the Venezuelan campaign, the U.S. parasitologist Charles Kofoid, and the distinguished German parasitologist Friederich Fülleborn. All agreed that hookworm control had impressively improved both the health of the peons and the profits of the United Fruit Company. The anemia of their laborers had abated, and with every additional gram of hemoglobin, there was a proportional rise of work efficiency and, with it, a proportional millions in profit.

Washburn, presenting the summarizing paper, brought the meeting to a close with a personal story that put the dollars of profit, the grams of hemoglobin gained, and the tonnage of feces rendered hookworm-egg-free into a human focus. In Trinidad he had an East Indian peon patient who suffered from severe hookworm disease. When this miserable Indian, by name of Seemungal, first presented himself at Washburn's clinic he was near naked, clad only in a skimpy breech cloth. It took four doses of drug to rid Seemungal of all his hookworms and after each treatment he donned more clothes—a hat after the first dose, trousers after the second dose, shirt and shoes after the third dose. And after the fourth dose, Seemungal felt so good that he changed his name to Mr. Christopher Padmore.

The religious-moral impulses of Gates, Rose, and even Rockefeller himself that spawned the Sanitary Commission and its global antihookworm programs had been transformed through the United Fruit Company into an enduring albeit too often neglected or forgotten economic principle: you make or get more money from healthy workers or taxpayers than from sick ones. By 1924 even the then president of the Rockefeller Foundation, George Vincent Todd, became an advocate of the economic benefits of good native health for the full exploitation of tropical dependencies.

This was 1924; in the moral tenor of those times colonialism was viewed as no evil thing. But the Rockefeller Foundation's health programs opened another, potentially more sinister door than that of colonial exploitation. It was the door of white colonization. The successes of the hookworm campaigns demonstrated that technology and strategies were available to make the disease-ridden tropics a more healthy, more livable environment—a place where the white folks might live in safety and comfort. Gorgas, the man who once looked forward to a benign old age surrounded by the happy black faces of his servants, spoke of the inevitable drift of the "white races" to the tropics once the health problems had been dealt with. Fortunately for the "brown races" those problems never have been resolved; in a paradoxically sick way, malaria, hookworm, yellow fever, and all the other torrid nasties have been the guardians of the tropical homelands.

The Rockefeller Foundation's international hookworm campaigns prompted the United States to look outward, to view the entire tropical world's medical problems. However, although it was the prime-moving parasite, it soon became apparent to the International Health Board that there was a host of medical troubles in the torrid zones—American and foreign. Moreover, these troubles could not be separated; they were commingled. Hookworms could not be separated from malaria; they were co-conspirators in causing the profound anemias of peoples living under conditions where both were highly endemic. The Rockefeller Foundation began to consider the entire spectrum of disease in the chronically ill tropics, but in this reconsideration it was a "new" acute disease that captured its immediate attention. Actually it was an "old" disease thought to have been brought under control. It was yellow fever. It had never gone away; there was more, much more, to that virus than had met Walter Reed's eye.

Chapter 9

Great Works 1915 to 1945: The Fever Is Broken

IN 1915, Wickliffe Rose, the apostle of global hookworm eradication, visited Malaya to negotiate for a Rockefeller-sponsored program in that British colony. The health authorities there agreed with him that hookworm was indeed a menace to public health although compared to malaria it was considered to be more of a nuisance than a killer. However, when the medical people in Malaya and throughout the Far East unburdened themselves of their most dreaded apprehensions, their worst nightmare was that of yellow fever coming to Asia.[29] When Rose returned he conveyed those fears to the acknowledged yellow fever meister, William Gorgas. Gorgas, fresh from his Panama triumph, was reassuring; not only could yellow fever be controlled, it

29. An epidemiological mystery that my colleagues and I have discussed over the years without coming to any conclusion is why yellow fever never came to Asia. All the right elements were there—the *Aedes aegypti* mosquito, the monkeys as potential zoonotic carriers, and, of course, the susceptible humans. One explanation is that there was no African slave trade to Asia and by the time extensive international travel was possible, an effective vaccine and a rigorous health screening of travelers from yellow fever countries was in place. Also, by that time steam had replaced sail, voyages were rapid and no barrels of drinking water to breed the Aedes were needed. Tom Monath, a tropical medicine expert who also worries about these historical lacunae, believes that dengue fever, caused by a virus related to yellow fever and present throughout Asia, might have offered some cross-protection.

could be "eradicated from the face of the earth in a reasonable time." Gorgas stimulated Rose to pursue yellow fever, as Stiles had done for hookworm a decade earlier. In 1916, the Rockefeller Foundation created the Yellow Fever Commission under Gorgas who, with his customary great energy, got down to work immediately by cleaning up Guayaquil, Ecuador. But Gorgas was not to see the new face of the disease to which he had devoted his life. On the fourth of July, 1920, in London while on his way to West Africa to establish a research station there, William Gorgas died of a heart attack.

When Gorgas died, great works had already been accomplished by the Rockefeller Foundation campaigns against yellow fever in the Americas. By hard sanitation work and by an abundance of hard money, in four years the Yellow Fever Commission had reduced yellow fever to the point of near-eradication in Brazil, Peru, Colombia, Venezuela, Ecuador, Panama, and the Caribbean islands. Only Mexico remained as the festering core, a menace as a source of reseeding the virus in its neighbors to the south and the north.

Americans may have mixed motives in their international charities, but we still feel hurt when no one says "thank you." It may be naive, but such is the way of our national character. We see only ingratitude where we should, instead, have empathy for the hurt national pride of the beneficiary countries, which may also have antipathy for the donor countries that they too often perceive as economic and political aggressors. So it was in Mexico in 1920; the Rockefeller Foundation's experience there is a timely lesson, worthy of closer examination, in the inextricable and sometimes conflicting interactions of medical and political diplomacies.

Mexicans hated Americans. And with good reason. Beginning in 1848, the United States ultimately appropriated 50 percent of Mexico. U.S. troops occupied Veracruz in 1914 and Ciudad Juarez in 1919. And if Mexicans detested Americans in general, they despised Rockefeller in particular. The very name Rockefeller was anathema; Rockefeller owned Standard Oil and Standard Oil was in total possession of Mexico's major source of wealth, its oil. Rockefeller the economic predator from Standard Oil and Rockefeller the philanthropist of his eponymous foundation were one and the same to the Mexicans.

For seven years, from 1913 to 1920, the Rockefeller Foundation's commission importuned the Mexican government to allow it to undertake an anti-yellow fever campaign. These proposals and offers of money and expertise were repeatedly rejected by the ardently nationalistic Mexican president, Venustiano Carranza. Then, in 1920 Carranza died and was succeeded as president by Alvaro Obregón. Obregón was also a nationalist, but he was a nationalist in trouble and he was willing to do business with the Rockefeller Foundation. Revolution against Obregón's government was flaring throughout Mexico, and troops were sent to deal with these insurgencies. The nonimmune soldiers entered endemic territory and were stricken with yellow fever. As a military-medical problem yellow fever was a very real threat to Obregón's shaky government. Now comes the unsavory business in which the Rockefeller Foundation, to gain a foothold in Mexico, is accused of making a deal worthy of Standard Oil.

In 1993, Armando Solorzano of the University of Utah published a paper in the *International Journal of Health Services* entitled "Sowing the Seeds of Neo-imperialism: The Rockefeller's Yellow Fever Campaign in Mexico." As the title implies it is a condemnatory account somewhat reminiscent of Marxist invective. But it is also a love-hate article that may well mirror not only Mr. Solorzano's feelings but also those of Mexicans in the whole historical Mexican-American relationship.

On the hateful side

Mr. Solorzano maintained that by the early 1920s Mexican physicians had yellow fever under control; no Rockefeller Foundation Commission was needed to show the Mexicans what they already knew how to do. Yellow fever was no longer a threat except to the government troops of a despotic regime. The real infectious killers were malaria, diarrheas, and pneumonias. Why wasn't the foundation pursuing these diseases if it was really interested in benefiting the Mexican people? The reasons says Mr. Solorzano (who got a grant from the Rockefeller Foundation to go through their archives in Tarrytown, New York) were: (1) Hubris. The Rockefeller Foundation sought only self-glorification; it wanted to be known as the savior of civilization from yellow fever. The health of the natives was of no real concern. (2) Imperialistic self-interest to advance the U.S. government's territorial ambitions and the economic designs of Standard Oil in Mexico. (3) Public health self-interest. Says Mr. Solorzano, the Rockefeller Foundation wanted to control yellow fever to protect the United States from being reinfected; it wasn't interested in Mexico as such. As proof, he asserts that the foundation's activities were to be limited to the port cities only, and so to create a *cordon sanitaire* around the Mexican coastline; the interior was to be disregarded. The foundation, the tool of the U.S. government, would pay for access by protecting the Mexican troops with a yellow fever vaccine developed by Hideyo Noguchi, a Rockefeller Institute scientist.

These odd assertions are a farrago of half-truths, the stirrings of a bygone animosity. In 1920, yellow fever was far from controlled by Mexican health workers. Aside from a few model cities, it more likely was in a trough of its natural epidemic cycle. Certainly, even then, there were numerous hot spots that remained uncontrolled. Wickliffe Rose wrote, "There can be no doubting the fact that Mérida is an endemic center of yellow fever, and as such it is a constant menace to Cuba and our own ports, to say nothing of the neighboring regions in México and Central America." The foundation replied to the criticism that it was neglecting the health problems of real importance to Mexico by reiterating its strategy of attacking one disease at a time to husband its resources. Also, yellow fever was not merely a major Mexican disease but an interregional one affecting almost every temperate and tropical country of the Western Hemisphere. Lastly, whether the Rockefeller Foundation pandered to the military's needs is a little unclear, but if it did, it wasn't doing the military any favors. The Noguchi vaccine, as we shall see shortly, was no vaccine at all but rather a useless, even dangerous horse serum preparation containing antibody to an organism totally unrelated to yellow fever.

On the love side

After the catharsis by impugning the Rockefeller Foundation's motives in Mexico, Mr. Solorzano comes to the post-campaign's epidemiological bottom line—love wins the day. It is impossible to reject the dramatic results. Within one year, the Rockefeller Foundation's Mexican Special Commission for the Eradication of Yellow Fever had reduced the yellow fever case and mortality rates by 80 percent! Two years later, in 1923, yellow fever was eradicated from Mexico.

It had been an enormous, labor-intensive undertaking. In 1921 there were no easy weapons, no insecticides sprayed by air or from ultralow volume-powered dispensers. There was only the Gorgasian operations of species sanitation throughout Mexico, the coast *and* the interior, house-to-house visits to clean up the *Aedes aegypti* breeding waters, give educational advice, and, where possible, to screen-in the houses.

And so, yellow fever vanished from Mexico and with it went much of the venom that had so long poisoned the relationship of the two neighbor countries. The campaign acted like a massive psychotherapist. The intimacy of the house-to-house visits and the constant display of caring and concern by the commission's workers changed the attitudes of the average Mexican toward the United States in a way that no grant-in-aid ever could. For the first time the United States was seen as a benefactor rather than a predator—a feeling that became embedded from the sacrifice by yet another martyr to yellow fever.

Dr. Howard Cross of the Rockefeller Foundation Mexican Commission specialized in bacteriology. In 1921 at Tuxtepac, he contracted yellow fever and died. His death deeply touched the hearts of the Mexican people, Catholics who appreciated saintly sacrifice. Cross was eulogized as a "martyr of medicine" and a "hero of humanity." Mexico gave him a state funeral in which he was mourned by all, "from the President of this great Mexico, whose heart beats true to every call of humanity, to the humble peon, who with bowed head reverently followed the funeral car."

The Rockefeller Foundation's yellow fever campaign by its effectiveness, its personal caring, and its giving up of one of its own ushered in the era of good (if not great) feeling between Mexico and the United States. The medical diplomacy was to have a far-reaching impact on political diplomacy and economics. In 1925, after the conclusion of the campaign, the highly satisfied director of the International Health Board, P. F. Russell wrote to his foundation president, "I have heard that there has been a real change in feelings in this country [Mexico]; . . . that before the popular feelings was pro-German and pro-English but that now while it cannot be called pro-U.S. it is inclined that way; almost all the automobile and merchandise one sees here is U.S. while before it was French, German, and English." Surely this is the handwriting on the wall of the North Atlantic Free Free Trade Agreement (NAFTA).

As in Havana, Panama, and Guayaquil, the cleansing of yellow fever throughout Mexico perpetuated complacency and the (erroneous) precepts of Walter Reed that the easily dealt with, domestic-loving *Aedes aegypti* was the only transmitting mosquito and humans were the only source of infection. Mexico was to be the end of the age of innocence; yellow fever would emerge as a much more complex and formidable disease than imagined by Gorgas. But before the biological facts became clear, there was a muddying of the pathogen.

In 1901 the Walter Reed group filtered the serum of an infected patient through a ceramic sieve whose porosity was so minuscule that no known microorganism could pass through it; and yet, when the filtrate of that serum was inoculated into two volunteers, both came down with yellow fever (fortunately, they survived). It was a partial, nevertheless convincing demonstration by Koch's postulate that the causative agent of yellow fever was a filterable infectious something too small to be seen under the microscope. In 1918, a brilliant, ill-fated Japanese microbiologist in the employ of the Rockefeller Institute challenged that viral (too-small-to-be-seen) assumption.

Hideyo Noguchi graduated from the Tokyo Medical College in 1887. Those turn-of-the-century years may well have been bacteriology's finest hour. Noguchi, swept up in the excitement of this new science, decided that he would become a bacteriologist. For his training he showed great independence in not going to Germany, as most other Japanese then did; instead he went to the University of Pennsylvania to study under the noted pathologist-microbiologist, Simon Flexner. In 1902 or 1903, when Flexner moved to New York City to become head of the new Rockefeller Institute, Noguchi went with him. He became a permanent member of the Rockefeller scientific staff, a new New Yorker, and a famed researcher who made significant contributions to the study of smallpox, syphilis, and snake venoms. However, it was with yellow fever that he came a-cropper, and his fame led many up the garden path of confusion. The phenomenon of fame and authority obliterating reason is all too familiar in the annals of science.

In 1915, two of Noguchi's bacteriology buddies in Japan, Drs. Inada and Ido, discovered a spirochete they named *Leptospira icterhohaemorrhagiae* which caused an acute, sometimes fatal disease marked by deep jaundice, now known as Weil's disease.[30] Because of the jaundice, Weil's disease pres-

30. We've been through syphilis, a spirochete-disease caused by a *Treponema;* and now there is this spirochete-disease caused by a *Leptospira.* Spirochetes are a large family of microorganisms; all are motile (by whiplike filaments called *flagella*) and are tightly coiled like corkscrews. There are three big groups (genera) that cause disease in humans: *Treponema* of syphilis and its relatives such as pinta and bejel, *Borrelia* causing Lyme disease and relapsing fever, and *Leptospira* of Weil's disease. Weil's disease is found throughout the world, but it is most common in the rural humid tropics, especially where there are livestock and rodents, animals that carry the disease. Infection is by ingestion of contaminated water; the "pristine" streams of Hawaii are a good source of leptospirosis for hikers, a recreational fact not well advertised in the tourist guidebooks.

ents itself, superficially, somewhat like yellow fever and the unwary may be trapped into misdiagnosis. That's what happened to the brilliant, misled Noguchi.

When one comes as a stranger to do research in a foreign country, it is usually necessary and may even be required to affiliate with the local medical people. It is they who determine who has what and furnish the material—blood, tissue samples, and so on—for your research. In 1918, Noguchi was sent by the Rockefeller Institute to Ecuador to study yellow fever. He was shown intensely yellow, acutely sick people whom the hospital doctors assured him were suffering from yellow fever. Noguchi inoculated guinea pigs with the blood he took from these patients. The guinea pigs became sick and jaundiced. When Noguchi examined the blood and tissues from these moribund animals, he was astounded when the microscope revealed a teeming field of leptospires. Noguchi in that instant became convinced that he had discovered what everyone else had missed, the true cause of yellow fever. It was a spirochete-leptospire *and* it was first recognized by the Japanese.

After renaming "his" spirochete *Leptospira icteroides*, he got down to the real work of producing a serum cure for yellow fever. He inoculated horses with the microorganism he had isolated in Ecuador, and in the serum from those horses he deduced were the antibodies that could cure patients of their yellow fever. Several thousand people in Mexico and El Salvador, stricken with yellow fever, were inoculated with the Noguchi horse serum. Noguchi discerned an effect, but when others reviewed his data, they found that those given the serum therapy were dying of yellow fever at the same rate as those not given the serum. Noguchi's reputation was so great that many of the scientists of the day were persuaded that he had, in fact, discovered the true etiological agent of yellow fever. Now doubts were beginning to fray the edges of his assertion and his reputation.

One who had doubts was another foreign physician-virologist transplant to the Rockefeller Foundation. Max Theiler came from South Africa to the London School of Hygiene and Tropical Medicine, then to Harvard Medical School, and then to the Rockefeller Foundation. Like Noguchi, he too was one of the greats of tropical medicine, a shrewd researcher who knew a red herring when he smelled one. In 1926, Theiler was at one of the most productive research laboratories ever established in the tropics, the Virus Research Institute in Yaba, a village on the outskirts of Lagos, Nigeria. He was trying to repeat Noguchi's work, but none of the guinea pigs he inoculated with yellow fever blood exhibited any sign of becoming infected. Theiler's rejection of the Leptospira etiology brought Noguchi to West Africa to see whether he could salvage his theory. In 1928 he went to Accra, Ghana, and there he contracted yellow fever and died. He was 52 years old. It is said that at the time of his death he, himself, no longer believed that his *Leptospira icteroides* was the cause of yellow fever. It is doubtful, however, that he ever realized that the ten years of research was squandered on the wrong microorganism, the pathogen from patients that he and the Ecuadorian doctors had mistakenly diagnosed as having yellow fever. Those patients had, in fact, Weil's disease.

It may not be biologically or even theologically proper to attribute human passions to lower forms of life; yet one can't help but feel that the wily yellow fever virus was consciously retaliating against its human adversaries. It laid down false trails to confuse its pursuers; it killed those who would dare hunt it; it led a secret, covert existence. Noguchi was not the last of the virus hunters to die, but those gallant men were beginning to reveal the secrets of yellow fever's epidemiology. One of those who made an important scientific contribution and paid the supreme sacrifice for doing so was Adrian Stokes.[31]

Stokes was a bacteriologist-physician, Irish, born in Switzerland, educated in Dublin, and when he went under the Rockefeller Foundation's auspices to the yellow fever laboratory at Yaba, professor of pathology on leave from the University of London. Among other aims of his research was to determine whether monkeys could serve as experimental models of yellow fever. Although there were all sorts of monkeys in West Africa, the institute had imported that faithful experimental human stand-in, the Indian rhesus monkey. Stokes inoculated a rhesus with blood of a Nigerian yellow fever patient. Within a few days the monkey became ill and showed all the symptoms of yellow fever—proving that monkeys, rhesus monkeys at least, were susceptible. Shortly after the monkey became ill, Stokes himself was stricken. The fever, jaundice, and debilitation all pointed to yellow fever. It was a mystery as to how he became infected. A stray wild mosquito? A stray laboratory-bred mosquito? An acci-

31. Between 1921 and 1930 six Rockefeller Foundation scientists, including Noguchi, Cross, and Stokes died of yellow fever. The last to be killed by the virus was a young doctor from South Carolina, Theodore Bevard Hayne. Hayne, a promising young researcher had joined the Rockefeller Foundation's laboratory in Yaba to help pay the debt of his medical education and because he loved field work. On July 6, 1930, he took ill with yellow fever and died, in respiratory failure, five days later. He was 32 years old; his wife in South Carolina was pregnant with a child Hayne would never see. Dr. Charles S. Bryan of the Department of Medicine at the University of South Carolina came across Hayne's tombstone in Congaree, South Carolina, and was prompted to write a touching tribute to this "Last Martyr of the Conquest of Yellow Fever" (*Southern Medical Journal,* 1993: 710–15). I can think of no other disease that killed so many scientists studying it.

dental needle stick inoculation of blood from the infected rhesus?

Stokes was dying. The character of this gallant man was such that even on the last days of life his thoughts were on his research and how his terrible illness could yet serve a useful purpose. He had shown that blood from a human patient could infect a monkey. The important question was whether a mosquito carrying the virus could infect a monkey, as would happen in nature. From his sickbed he directed that *Aedes aegypti* mosquitoes from the institute's insectory be fed on his jaundiced, fever-racked body and that those mosquitoes, after a suitable incubation period, be fed on a clean rhesus. This was done and the man-mosquito-monkey transmission was accomplished, but when that happened Adrian Stokes was dead of yellow fever. This was in 1927 when Stokes was 40 years old.

The suspicions bred in the Yaba laboratory that monkeys might somehow be involved in the natural history of yellow fever began to be confirmed after the death of Stokes. The Rockefeller Foundation had expanded its yellow fever effort by creating and staffing research posts spanning Africa from Lagos in Nigeria to Entebbe in Uganda. In South America there were Rockefeller laboratories in Rio de Janeiro and Iléhus, Brazil, and Bogotá and Villavicencio, Colombia. Wild monkeys were trapped and sent to these laboratories where their blood was searched for the presence of either the yellow fever virus itself or its "footprint," the specific antibody raised against it. The first indication of the monkey-virus connection came from the Ilhéus laboratory when they found a forlorn, sick marmoset monkey in one of their traps. The animal died a few days later, and the virus isolated from its blood proved that it had succumbed to yellow fever. Now an exhaustive monkey hunt was mounted in tropical Africa and South

America. The results stunned the investigators; either the virus or the sign of the virus, the antibody to it, was detected in a large percentage of monkeys. In some troops 70 percent or more were positive.

This upset the old epidemiological apple cart. First, it signified that yellow fever was not entirely a human infection concentrated in urban centers. The monkey findings indicated an enormous reservoir of infection in the tropical forests—a sylvatic cycle. Second, if there was a sylvatic cycle of transmission, then it must be by mosquitoes other than *Aedes aegypti.* And third, transmission from monkey to man must be taking place, which would explain the focal outbreaks in remote forest or forest-bordering villages.

During the 1930s and 1940s a beautiful series of field and laboratory researches, largely sponsored or undertaken by the Rockefeller Foundation, in Africa and the tropical Americas proved the above suppositions to be all too true. In both regions, species of mosquitoes other than *Aedes aegypti* were identified as vectors in the highly complex cycle of sylvatic yellow fever. In most circumstances it would be best to leave it go at that; to note that it is all terribly complex—a tangle of behavioral-ecological-vertebrate-host–invertebrate vector factors peculiar to each endemic setting. However, we are living in an ecologically [illegible] and destructive time and are paying the price. "New" infectious diseases seem to appear from out of nowhere, and we are bewildered. Of course there is no nowhere, no spontaneous generation. Infectious diseases often come to us from the animal reservoir when that complex, delicately interacting epidemiological machinery is upset. Yellow fever serves as an excellent model of that complexity and it would help us to examine at least one setting where it descends from the trees, from monkey to man. Let us go on a yellow fever safari to East Africa.

The black mangaby monkey lives high in the tree canopy. It's a social homebody; packs of the animals have been observed to sleep each night in the same tree for over five years. Also confined to the forest canopy is the high-flying (acrodendrophilic, as the purists would term this behavior) *Aedes africanus* mosquito. This mosquito is an efficient carrier of the yellow fever virus; 70 percent or more of the mangabys have been found to be infected. However, the virus seems to be benign in the mangaby; they act as a reservoir, a sort of Typhoid Mary monkey. Also sleeping high in the trees is a neighbor monkey of the mangaby, the redtail monkey. These primates become infected with the virus via the mangaby–*Aedes africanus* route. They too are yellow fever Typhoid Mary monkeys.

All this would be of academic interest if the redtail would stay high in the trees like the shy mangaby, but it doesn't. The redtail is a bold creature and comes down to raid the gardens and plantations at the forest edge (the ecotone). Here in the clearings at the forest edge is another aedine, *Aedes simpsoni,* also an efficient carrier of the yellow fever virus. As humans have destroyed more and more of the forest and settled in these clearings, *Aedes simpsoni* has become more and more domesticated, biting human settlers and the redtail monkey marauders indiscriminately. So now the viral transmission lines have been extended from the redtail monkey through *Aedes simpsoni* to the human.

The infected human carriers (the sylvatic virus cycled through monkeys seems to be less virulent than the urban form which seems to hot-up when passed successively from human to human) go to town to market their crops. In town, the country human is bitten by the urban *Aedes aegypti,* completing the familiar urban cycle of yellow fever of human to human via *aegypti.* With human intrusion and degradation of natural ecosystems, one doesn't have to be a a medical entomologist to visualize the behavioral and species population changes that would make yellow fever or any other lurking zoonotic infection a threat to human health.

With the revelation of the sylvatic cycle in the 1930s and 1940s, the hopes for the African–tropical American eradication of yellow fever collapsed. Even the Gorgas strategy of urban control by species sanitation of *Aedes aegypti* was proving, in the long term, to be impractical; impoverished city dwellers in tropical countries simply did not have the will to maintain antimosquito measures. Keep in mind that this was before DDT and the quick fix that it made possible. However, the decades of the 1930s and 1940s saw the beginnings of modern microbiology and with them the prospects of a new, powerful, and practical form of prevention—a vaccine to protect against infection with the yellow fever virus. The yellow fever filterable "particle" had been identified as a virus. Methods had been established to propagate it in the embryos of chicks within their eggs and in a convenient experimental animal, the mouse, by inoculating directly into its brain (not a technique for the squeamish). Once again, it was the Rockefeller Foundation that took up the challenge and pioneered the way to the vaccine.

Times were a changing at the Rockefeller Foundation. In 1923, Wickliffe Rose, a man motivated by the morality of old-time religion, stepped down as head of the International Health Board. The man who succeeded him was of a different temperament and philosophy. Colonel Frederick F. Russell was a trained, dedicated physician-scientist who had been chief of the U.S. Army Surgeon General's Office's Division of Laboratories and Infectious Disease. Russell believed in the seamless inseparability of the Rockefeller Institute and the Rockefeller Foundation's International Health Board. The institute through its research was to discover new avenues for world health, and the board was to be the vehicle that traveled down those avenues. Russell grasped the reality that to deal with yellow fever, a vaccine had to be developed and the institute was the facility best suited to carry out the research. The man chosen to lead that research was another physician-microbiologist, Dr. Wilbur A. Sawyer.

Sawyer's team began work in 1931 by following the Pasteurian path of vaccine development by inactivation of the pathogen. However, rather than drying out the spinal cords of infected rabbits, as Pasteur had done to make his rabies vaccine, Sawyer took the alternative tack of trying to reduce the yellow fever virus's virulence, while retaining its immunogenicity, by passaging it successively through a series of different animals. In a long series of tedious experiments, the virus was taken from the human patient, given to the rhesus monkey, and in turn, given to the mouse. Through this procession the virus remained adamant; steadfast in its virulence, it killed experimental mouse after experimental mouse. The next step was first to incubate the virus in the serum of humans who had recovered from yellow fever in the hope that the antibody in the serum would modify the virus. It helped, but again, passage after passage the virus still remained too virulent to be used as a vaccine in humans.

During the course of this line of investigation the virus again took its revenge on its pursuers; Sawyer and six of his team, despite what was then the most stringent of laboratory precautions, somehow became infected. Fortunately, all survived, but it shook the Rockefeller administration and Russell decided that it wasn't worth the sacrifice of his best and brightest scientists. He was about to order the yellow fever team to close shop, but Sawyer persuaded him to stay the course and the research continued.

The group's next experimental wrinkle was to do away with the middleman, to delete the animal passages and go from the viral isolates through a series of tissue cultures, while inactivating the virus with immune, antibodied serum at each passage. It worked! The virus was rendered nonvirulent and was effective as a vaccine protecting experimental animals. However, it was effective as a vaccine only when serum was given as a component of the immunizing package; no serum, no immunity. It was, nevertheless, a major breakthrough and it was good enough for the Rockefeller Foundation. Pilot trials were begun, but it soon became apparent that this was not the vaccine that the yellow fever endemic regions of the world had been waiting for. In the 1930s there was as yet no awareness of the silent contaminants, notably the hepatitis viruses (and, of course, today the HIV virus would be the prime example) in blood that was to be used for transfusions or for biological purposes such as when admixed in vaccine formulation. In the wake of the pilot trial there were cases of hepatitis and that ended the prospects of the serum-inactivated yellow fever virus vaccine. In 1936 Sawyer and company had come to the end of their line; they had shot their experimental bolt. They didn't know where to go next; they needed a miracle. Miracles are unexpected and, well, they are miraculous. For Sawyer it came in the guise of an illiterate West African peasant.

In 1927, a 27-year-old man named Asibi was ill with yellow fever. His worried family, after giving him the medical benefits afforded by the village juju man, brought him to the hospital in Dakar, Senegal. Blood was drawn from Asibi and sent to the Virus Research Laboratory in Yaba where it was inoculated into a rhesus monkey newly arrived from India. On July 4, 1927, the monkey died of yellow fever. Happily, Asibi lived. Over the next nine years this yellow fever virus isolate, the Asibi strain, was successively passaged through mice, chicken embryos, and monkeys by Sawyer's people at the Rockefeller Institute in New York City. All that while it remained as deadly as the day it killed the rhesus in Yaba. Sometime in 1936 during a passage in the chicken embryo the miracle happened; the virus mutated, and lost or altered the still mysterious gene that confers the property of virulence. The mutated virus was alive and it was infectious; it just didn't make mouse, monkey, or eventually, human ill. What it *did* do was elicit protective antibody. Inoculation with the 17D strain, as it was designated, made the volunteers from the Rockefeller laboratory immune to challenge with a hot strain. Asibi-17D was the ideal vaccine that Sawyer had hoped for during the six long years of research. It was highly immunogenic, and importantly, it seemed stable and didn't revert to its old, nasty, lethal ways.

Today it would require at least five years, $20 million, and 20 cartons of documentation to bring a candidate vaccine to a phase III (an extensive, population-based) trial. One looks backward in amazement at the way the Rockefeller seized the moment. Within *six months* of 17D's emergence from the chicken embryo in which it had mutated, the Rockefeller Foundation had conducted its safety and efficacy trials, gone into full scale vaccine production, and begun its vaccination program by immunizing 40,000 Brazilians and Colombians. Even Pan American Airlines had its flight personnel immunized. It was also a time when the German evil began to threaten the democracies. The military medical people, who were often more politically wise than their politician masters, knew that war was inevitable. Now they had the 17D vaccine to protect troops that would be deployed in tropical Africa and South and Central Americas. From 1942 to 1946 the Rockefeller Foundation produced 34 million doses of the

17D yellow fever vaccine. All was given free to government health agencies throughout the endemic world. In 1946, Rockefeller closed its vaccine production and handed over the manufacture to national governments, again completely without charge. It was an act of generosity, of great philanthropy, that has never been equaled or repeated by others, private or government.

Also never repeated was the miracle of that mutation; another nonvirulent, vaccine-capable strain of yellow fever never again emerged from the laboratory. The gift of Asibi continues to protect all of us who venture into the torrid zones. Its use in native populations at risk is still limited, but it can be used in epidemic flare-ups. With yellow fever it's not over till it's over; the vast sylvatic reservoir remains, and outbreaks, although not on the grand scale of yesteryear, still occur. For the moment, the virus is contained and the scientists who gave their lives in the yellow fever wars have been rewarded.

For the Rockefeller Foundation, medical missionaries seeking new diseases to conquer and new worlds to heal, it was time to turn to the greatest killer of humans at home and abroad—the mother of fevers, malaria!

Chapter 10

Great Expectations 1900 to 1945: Malaria—Death at Our Doorstep

IN THE SUMMER of 1926 a prosperous looking Yankee came to a town in the Carolinas. The mayor thought the man an agent of industrialists up north who was scouting for a place to locate a factory, a textile mill maybe. This is a great town, the mayor tells the man. It's as healthy and happy a place as you can find anywhere in the United States. We've got lots of good workers for you, real go-getters, full of energy. And they work cheap. Cheap! Especially the blacks, they'll work practically for nothing. Bring your kids here; they'll grow up big and strong.

The man corrects the mayor. He has mistaken his identity. He's not an industrialist from up north but a government agent working for the Public Health Service who has been sent to investigate health problems in the region. On hearing this, the mayor turns on the man; he is livid with anger.

Investigate! Investigate! That's all you damn government people can do! This place is a pesthole, a deathtrap. We got so much malaria fever, everybody's so sick and weak all the time they can't work a lick. The malaria is taking our blood. We had a mill here but nobody could come to work regularly; they was too sick. The mill had to close and now we haven't any proper employment. Spring planting went to hell because half the farmers were down with the fever. We need help. For God's sake man don't just investigate, help us. We need help. Do something!

Imagine yourself a farmer, the owner of 15 acres of good land about a mile from the distraught mayor's town. This early summer morning is a glory of the Carolinas, an intense blue sky, the musty smell of newly plowed sandy earth mixed with a faint resinous hint from the stands of longleaf pine. Crows wheel and call. On such a day you might look up from behind the old mule as it knowledgeably pulls the plow in a straight furrow, marvel at the beauty of the day, and have a sense of contentment, even happiness. But today you know you're in for trouble. The fever is about to come.

It begins in the way you know all too well since childhood—a shadow, a painful throb in the head. A chill comes over your body that denies the warmth of this southern season. You know that you should unhitch the mule and ride it into town to see the doctor. You did that last spring when the malaria came on strong. The doctor charged $2 and another $2 for the medicine; quinine he called it. The quinine broke the fever. You could work two days later, but it certainly made your head sing; that high pitch twanging still never lets up in your ears. That $4 was money well spent, even though you only make about $400 a year from the farm and from that there's six mouths to feed, six bodies to clothe, and supplies to buy for the kids' schooling. This spring there was a lot of rain and the cotton bolls will be fewer and smaller; you'll be lucky to clear $300 from the crop. There's no money to spare for medicine except maybe for the store-bought medicine for your wife. She's got "woman's troubles" and that tonic makes her feel better even though she's kind of dreamy most of the time since she's been taking it. The rain brought a lot of mosquitoes. The doctor says that the mosquitoes bring the malaria, but your old daddy used to say that the fever came from the bad air coming up from the wetland. Now you're getting cold and shaky. It's the scattergue. You better get to the house and go to bed.

You are shaking uncontrollably as you stagger into the house. It's the deep cold in your body; never on a winter morning have you felt such an icy bite in your marrow. Your wife knows the trouble; she knows what to do. In bed she covers you with every quilt, every blanket. A roaring fire is lit and still you shiver. The headache is now like a searing iron in your brain; your bones ache. There seems to be no let up; the sickness goes on and on although your wife later tells you that you had the shakes for about three hours.

For a moment there comes a relief; the shivering stops. The respite is brief and now the heat comes; your body burns with an intense fever. The covers are thrown off. The sweat pours off you like a river drenching your clothes. The head throbs with an even greater pain and you can't think too clearly; it's like a drunken delirium and vicious hangover all together. The hot phase lasts for about four hours and then, mercifully, the great fever breaks. You are so tired, so baby weak. Exhausted, you sleep.

The next morning the fever is still gone. You feel a bit better but are still too weak to even think about working. The oldest boy has stayed home and is trying to do the chores. But he's only 11, and you can't send a boy to do a man's job, no matter how willing the boy. You also know from those past springs and summers of malaria that it is not over; the devil in your blood had not yet run its course. The expected new attack comes two days after that first rigor. It goes through the all-too-familiar course, not quite as bad as that first rigor-sweats but still bad enough, and this time you feel nauseous, you vomit, and your bowels loosen. Another two days, another attack, and then it's over; the malaria has finished with you for this season.

That next week you move listlessly about the house, sit on the porch with your face in the sun. That's about all you can do. Except worry. The thought of poverty, the inability to provide for your family, is the constant dark companion of your mind. In other years when the crop was poor or sickness struck, you could make it by working extra in the lumber mill in town, but the mill closed last year. There was so much malaria sickness that the mill was almost always short handed and the owners, from up in Boston, couldn't make a profit, so they closed it down. How grimly funny; so many people in the County out of work and there wasn't enough healthy labor to keep the mill going.

And it is here that our imaginary narrative stops. No happy ending. No resolution. I don't know what you, the poor Carolina farmer, do next in the aftermath of your malaria.

History and statistics. In 1919, the U.S. Public Health Service summed up malaria in the U.S. South with the statement, "for the South as a whole it is safe to say that typhoid fever, dysentery, pellagra, and tuberculosis, all *together*, are not as important as malaria." This could equally be the assessment of 200 malarious years throughout most of continental America. By 1650 malaria was well established in Connecticut. In 1680, Charleston, a young city surrounded by rice plantations—nurseries of malaria-carrying mosquitoes—was almost abandoned because of the intensity of malaria. When the Revolution erupted in 1776, the Carolinas were the most malarious region in North America. As the country pushed westward, malaria went with advancing frontiers. When the United States entered the twentieth century, malaria thrived from New York to Florida. Malaria reigned in Kansas, Texas,

Oklahoma, the Ohio Valley, the Mississippi valley, and the Sacramento and San Joaquin valleys of California.

During the first quarter of the twentieth century, an estimated 5 to 7 million cases of malaria occurred in the United States *each year;* from 1914 to 1923 malaria caused approximately 10,000 deaths. In 1919, Savannah, Georgia, with a population of 10,313 recorded 2,476 cases of malaria. Surveys carried out in North Carolina from 1910 to 1920 revealed that in Pamlico County 32 percent of the whites and 38 percent of the African Americans had malaria parasites in their blood, in Beaufort 50 percent were infected, and in Roanoke Rapids 75 percent. This roll call of malaria in the United States could continue but would serve no further purpose in the statistical exhibit of this so very American "tropical" disease. What we should not lose sight of in these impersonal statistics is the fact that each case, every number, represents a very sick, sometimes fatally sick, person. Our Carolina farmer had a relatively moderate bout. Within the statistics would be the men, women, and children chronically incapacitated by malarial anemia, there would be women dying in childbirth of acute malaria of pregnancy, there would be comatose children dying of cerebral malaria, there would be children with 105 or 106°F fever burning to death. And keep in mind that this was the United States in 1926, not in pre-Revolutionary 1726. The burden of malaria fell most heavily on the South where it frustrated progress and prosperity.

It may not seem like it to younger (under 40 years of age) readers, but 1938 was modern times. Automobiles were beginning to look streamlined; airlines were criss-crossing the country; television was in the works; Benny Goodman, Tommy Dorsey, and Glenn Miller were swinging. The country was beginning to emerge from the Great Depression although the continuing economic retardation of the South was of concern to President Franklin D. Roosevelt. He assembled a National Emergency Council of experts and in 1938 it submitted its "Report to the President on the Economic Conditions in the South" which accused malaria of being a major economic retardant. The report stated:

> The pressure of malaria which infects annually more than 2,000,000 people, is estimated to have reduced the industrial output of the South one-third. In reports obtained in 1935 from 9 lumber companies, owning 14 sawmill villages in 5 Southern states, there was agreement that malaria was an important and increasing problem among the employees. During the year 7.6 percent of hospital admissions, 16.4 percent of physician calls, and 19.7 percent of dispensary drugs were for malaria. . . . Ten railroads in the South listed malaria as an economic problem and costly liability. . . . Each case of malaria was said to cost the companies $40. In we attempt to place a monetary value on malaria by accepting the figure of $10,000 as the value of an average life and using the death rate of 3.943 for malaria reported by the census of 1936, the annual cost of death from this disease is $39,500,000. To this figure could be added the cost of illness, including days of lost work.

The lost time from work caused by malaria averaged, from state to state, from two to four weeks for each worker each year. Those who returned from the malaria sickbed were in no shape to be highly productive; they would not be the people to build BMWs or Hondas or, in those years, to run machinery in a textile mill. Southern administrators and politicians realized that modern times were being subverted by malaria. They were desperately trying to attract capital, to have northern industries relocate to their states. However, malaria and other southern illnesses were not under effective control, so the regional politicians adopted the simple strategy of dissimulation. They told the capitalists that all was well, that the regional health problems were either being dealt with or nonexistent. Brewton, a town in Alabama, hung a banner across its Main Street declaring "Malaria Being

Controlled—Come Locate With Us For Health and Prosperity." The prickly, defensive, sensitivity of the malarious South was even to appear in the sports section of an Alabama newspaper. In 1926 the University of Alabama's Crimson Tide made their first ever trip to the Rose Bowl where they defeated a highly favored Washington University team. The *Montgomery Advertiser* reported the triumph with, "The feat of the Crimson Tide should go far to dissipate certain popular illusion concerning hookworm and malaria in the South." In 1926 the malaria death rate in Montgomery is estimated at 90 per 100,000. Words of concealment were patently not an effective antimalarial strategy. But words were cheap and nothing much else was being done.

Ignorance was no excuse. By the turn of the century the essential knowledge needed to design an antimalaria campaign was in place. Alphonse Laveran, the French army doctor in Tunisia, had discovered the malaria parasite within the red blood cell in 1880. Ronald Ross, the British army doctor in India, had discovered the developmental cycle of the parasite, leading to transmission, within the mosquito in 1898. Later in 1898, Giovanni Grassi, the Italian doctor-biologist, discovered that only the Anopheles mosquitoes transmitted the malaria parasites of humans. So, when the world welcomed the twentieth century, it was known what caused malaria and how it was transmitted by the bite of the Anopheles mosquito. Admittedly, no long-acting insecticides or cheap, effective antimalarial chemotherapeutics and chemoprophylactics for mass distribution existed, but it was well understood by most state health authorities (except Florida which stubbornly clung to the miasma theory of malaria transmission until 1902) that getting rid of the anophelines or even keeping them from biting humans would prevent malaria. They knew that the main U.S. vector, *Anopheles quadrimaculatus,* bred in ponds and swamps as well as in smaller water collections such as inundated cornfield furrows. They knew that engineering-sanitation works could deny the female "Quads" their essential breeding waters. Two antimosquito operations in the North showed the way.

High above Cayuga's waters was a very malarious place. With the coming of spring and temperatures rising to 70°F, thousands of malaria cases erupted each year in Ithaca, New York. In 1905 the city fathers of Ithaca committed funds to begin a large-scale antimosquito program. The nearby wetlands were drained and the remaining standing waters were oiled, so the aquatic mosquito larvae who breathed air through their siphons were suffocated. Three years later, in 1908, Ithaca was malaria free and has remained so.

In 1900 one in every five residents of Staten Island had malaria. The island was under the responsibility of the Port of New York Authority, and in 1901 one of its health officers, Dr. Alvah Doty, proposed that an antimosquito campaign be undertaken. He was given $50,000—a lot of money in those days—to do the job, and with those funds Doty constructed a series of canals to drain the marshes surrounding much of the island. By 1908, Staten Island, like Ithaca, was malaria-free. The island then prospered as a residential haven from New York City; property values soared, and with them, the property tax revenues. That $50,000 had been the best of investments.

Unfortunately, few southern municipalities or counties had the money to undertake those antimalarial operations during the first decades of the twentieth century. Repeated appeals for help went out to the still-new U.S. Public Health Service. The Public Health Service was of little assistance; it could give sound advice but no money. The federal government didn't allocate a budget for its public health service to carry out public health operations. In 1915, the federal government showed its disregard, if not disdain, for the well-being of its southern citizens by allocating $3,000 for antimalaria activities. That was for the *entire* South. But then, as now, when the military deemed that *it* was at risk, no expenditure was too great. In 1918–1919 an antimalaria budget of $580,000 was voted that was exclusively devoted to the protection of World War I soldiers training in southern camps.

In desperation, the U.S. Public Health Service canvased the business community for antimalaria money, but drew little response to its requests. The southern business community had mobilized to fight yellow fever when their state and national governments defaulted. Yellow fever was a disease with attention-getting impact; it struck as a commerce-paralyzing epidemic. It had the drama and sudden effect that the business community responded to because it perceived that it was in its self-interest to do so. Malaria was chronic, so much more insidious than yellow fever. It was almost an accepted way of life in the South, and the business community didn't respond despite the abundant evidence that malaria was *the* major economic health liability.

In the Third World and the colonial world from which it devolved, it is the missionaries who often provide the health and education services that the governments cannot or will not give. For the Third World that was the U.S. South during the pre-World War II twentieth century, the Rockefeller Foundation was the missionary. However, for all its zeal and good works, the foundation was late in coming to the problem of malaria. The leaders of the foundation in that era were essentially religious professionals, not public health professionals or scientists. They followed the advice of their prophets and there was no Moses of malaria as Stiles had been the

Moses of hookworm and Gorgas the Moses of yellow fever. When malaria caught the foundation's attention, it did so in a relatively minor key, but even so the effect was great for a limited expenditure.

In 1914 Wickliffe Rose met with Ronald Ross in England and Malcolm Watson in Malaya. Both men impressed on Rose the importance of malaria not only globally but also in his own country. Although Rose became convinced, his commitments, energies, and funds were with the hookworm. In 1916, malaria got a token $50,000 from the foundation. The foundation made the best use of this money by entering into a partnership with the U.S. Public Health Service and four local communities to establish pilot projects that successfully demonstrated how malaria could be controlled by antimosquito environmental engineering. More money was given (but the total for *all* malaria activities, including those in other countries after the foundation went global in 1923, amounted to only about $5,000,000 over the 39 years that the foundation pursued malaria). Research field stations were established where important studies were carried out on the epidemiology, entomology, and control of malaria. By 1922, the Rockefeller Foundation operated in 163 counties of 10 states. Most of their activities were in urban centers. It was a great limited success; by 1920 malaria had virtually disappeared from the cities of the South. It had cost about 45¢ a head to do so. However, the countryside and its large rural population remained as malarious as ever. More than the Rockefeller Foundation would be required to finally free the United States from malaria.

The eradication of malaria in the South became an element in the web of a shifting political philosophy on the role and responsibilities of the national government. It was a byproduct of the conflict between the Roosevelt federal administration and the private utilities. The federal government first edged into the power business in a legitimate, nonthreatening way. During World War I the military needed a facility to manufacture gunpowder. To make gunpowder on a wartime scale, you need lots of electric power, and in 1917 no private power company could supply that kind of wattage. The feds, taking the matter into their own hands, built a dam at Muscle Shoals, Alabama, from the dam they built a hydroelectric generating plant, and by the hydroelectric plant they built a gunpowder plant. This very efficient operation produced electricity at a much lower cost than that from the private utilities. The spare kilowatts went to homes and factories in the surrounding region which prospered thereby. The private utility companies, which had done so little in the way of rural electrification, feeling rightfully threatened, lobbied that the government should get out of the power business at the end of the war—which it did. Nevertheless, Muscle Shoals was a warning to the private utilities; it was a demonstration of what the federal government could efficiently do in the way of grand water impoundment–hydroelectric schemes.

There had been a progressive ecological degradation in the major southern river basins, notably the Mississippi and Tennessee, that was near equal to our era's assault on the Amazon rain forest. The land was deforested; erosion followed and uncontrolled floods followed that. In 1927 there was a great deluge in the Mississippi basin. Farms and farmers were swept away, and a watery wasteland left. President Calvin Coolidge sent the engineer Herbert Hoover to inspect and report. Hoover recommended that the government construct a series of levees and dams with power-generating capability upstream to control the flooding and bring the region back to life. Coolidge was silent in reacting to Hoover's report. Nothing was done. There was a great amount of standing water after the floods had receded, a breeding delight for *Anopheles quadrimaculatus.* The years between the great deluge and the Great Depression were one of the most intensely malarious periods in the U.S. South.

Franklin Roosevelt was elected in 1932. He came to lead a nation deeply wounded by the poverty of the Great Depression and with its land degraded from ecological abuse. Ill health was the shadow cast by this very bad time; malaria and malnutrition were sapping the strength and resolve of the South's rural people. Roosevelt seized on the idea that a Great Work in the South would galvanize the country and give promise where there was despair. The Hoover Report was the ready-made plan on which Roosevelt could act.

The man who stood with Roosevelt was the senator from Nebraska, George William Norris. Norris was either, according to one's political ideology, a fearless liberal Republican or a renegade who betrayed his party's principles (and was ultimately expelled from the GOP to become a repeatedly reelected independent). Norris, the representative from an agricultural state, was concerned with the plight of the U.S. farmer. He too envisioned a Great Work to bring the rampant rivers under control, irrigation to the fields, and electricity at an affordable rate into the rural home. It would demonstrate by "creeping socialism," as Dwight D. Eisenhower was to later characterize the Tennessee Valley Authority (TVA), how the power and resources of the federal government could fill the needs of a people that Roosevelt-Norris considered to have been abandoned by the private sector. The TVA was to be that Great Work, and in 1933 Norris fathered bills to create the TVA. Neither Roosevelt nor Norris foresaw that from the dams and the wattage there would also be the beginning of the end of that most American of diseases—malaria.

When the TVA began to construct its series of dams and drainages in 1933, an estimated 150,000 cases of malaria resulting in 5,000 deaths annually occurred in the Tennessee River basin. Surveys revealed that as many as 65 percent of the people had malaria parasites in their blood. That is a level of malaria seen today in only the most hyperendemic settings of tropical Africa or Papua New Guinea. As bad as the pre-TVA malaria was, the expectation was that it would further increase once the project got underway. Water reservoirs had been built elsewhere in the South from 1910 to 1926, and in almost every instance they brought more mosquitoes and more malaria.

To their very great credit, the TVA directors were sensitive to the danger, and even before the first dam was thrown across the river, an expert team of malariologists, entomologists, biologists, and engineers was hired to become a permanent unit and to figure out ways by which mosquitoes and the malaria they carried could be controlled. The greatly increased expanse of impounded water, 21 dams behind which lay a total of 600,000 acres of lakes were built along the system, made an immense new watery bed for the "Quads" to breed. In those early TVA years malaria rates increased. The entomologists discovered that only the margins of the lakes were used by the females to lay eggs and the larvae congregated around the aquatic foliage in the shallow waters. The entomologists-malariologists and engineers, working together, figured out that by strategically raising and then lowering the water levels, which now could be controlled at will, the *Anopheles quadrimaculatus* larvae could be dried out and thus killed. Drainage of the wetland swamp waters also reduced the number of mosquitoes. Numerous canals were built which made the water flow faster, and that too destroyed a large number of "Quad" larvae. Another strategy, clearing the aquatic vegetation, denied larvae of habitat and also made the water flow at a greater velocity in the canals.

Malaria rates began to decline, but slowly. As late as 1942 there were still about 50,000 cases of malaria annually although better management and improved health services had reduced the mortality to "only" about 600 people. The number of *Anopheles quadrimaculatus* was still high enough to allow for continuing transmission of malaria in the Tennessee River valley. Something else was needed to put the finishing touch to the TVA antimalarial operations. The knockout punch was the much-maligned DDT.

I have defended DDT's medical application elsewhere (in *The Malaria Capers*). Suffice it to reiterate that DDT was one terrific weapon against mosquitoes and other insects of medical importance. It may well be that the next-best thing after the defeat of the Nazis to have come from World War II was the residual insecticide DDT. In the 1930s the TVA tried to control mosquito breeding by casting kerosene and/or Paris Green (a powdered arsenical formulation) on the waters, without much success. DDT came to the United States from the Swiss pharmaceutical firm Geigy which, in 1940, passed on the compound and method of manufacture to the U.S. Army military attaché who passed it on to Washington. Tests quickly showed that it was the most powerful of insecticides ever developed and had the unique character of being long lasting, having a residual effect. It saved countless lives in Europe among the wretched refugees who were DDT "dusted" to kill the typhus-carrying body lice.

The TVA entomologists were remarkably quick to

appreciate DDT's potential. In 1943 the TVA had planes spraying the Tennessee River valley with DDT, like so many crop dusters. The spraying went on each late spring to early summer, the height of the mosquito breeding season, for almost ten years. Entomology teams in their boats patrolled the drainage canals and lakes casting DDT into the water to kill the larvae. About the same time, the herbicide 2,4-D was discovered and, like a dress rehearsal for Vietnam, it was liberally put into the TVA system to kill the aquatic plants the "Quad" needed in their ecological nursery. The malaria rate plummeted; in two years, by 1947, a negligible 2,000 or so people had a malarial attack. Meanwhile, the malariologist's canals became waterways for recreational boats, the dried swamps became refuges for waterfowl and pasturage, and the lakes became recreational sites for thousands of Americans.

However, throughout the other parts of the South malaria continued to take its toll, particularly on the poorly immune children and the first-time pregnant (primigravida) who lose their immunity. In 1941, modern times, a Dr. Richard Torpin of Augusta, Georgia, described in the *American Journal of Obstetrics and Gynecology* his series of 27 cases of "Malaria Complicating Pregnancy." Those cases, women in the United States of America, could be matched then and now in every detail by African and Asian women stricken with severe malaria of pregnancy. Here are Torpin's synopses of five of his patients:

1. White woman, aged 20 years, at term, estivo-autumnal malaria [an old name for *Plasmodium falciparum* malaria, given because it mostly came in late summer and early autumn], comatose, died following delivery of stillborn infant.
2. Gravida, 6½ months, stillborn, labor before quinine given.

3. Gravida, 2 months, began to abort prior to administration of quinine.
4. Gravida 4 months, aborted before treatment began.
5. Gravida, 8½ months, premature delivery during an attack of estivo-autumnal malaria before quinine given.

One severe case had a good outcome, but her history illustrates the trials of the pregnant southern woman during the first half of this century.

White woman, aged 18 years, primigravida, became ill with estivo-autumnal malaria at three months' pregnancy and for four months she was recorded to have chills and fever daily. She had headache, dizziness dyspnea, fainting attacks, and became extremely anemic, the lowest recorded red blood count being 1,090,000 [7,000,000 normal] and hemoglobin 10 per cent [16 per cent normal]. She lost weight to 87 pounds. She was treated with repeated transfusions totaling 3,000 c.c. and an aggregate of 350 gr. of quinine. In spite of all this and two attempts to induce labor by castor oil and strychnine she continued pregnant to term, when she spontaneously gave birth to a healthy live baby weighing 7 pounds 2 ounces.

Other World War II–associated factors also contributed to malaria's disappearance. The TVA health educators had, since 1933, been making house-to-house rounds preaching the antimalaria gospel of prevention—screen your houses, don't go out at night when the anophelines bite. The Tennessee River valley people, as well as most of the other southern rural people, were almost always too poor to act on that good advice. With the TVA, crop yields increased, and cheap electricity gave further opportunity for employment and enriched the life of the farmer. World War II finally lifted the United States from its economic doldrums. There were jobs and there was money. And money was, as it still is, the best antimalarial. Houses were screened in, the mosquitoes excluded.

People slept better and they didn't get malaria. By 1952, malaria, which had been the torment of North Americans for 300 years, became a forgotten disease, a curiosity even to medical students.

There is a present-day postscript to this. Malaria in the United States may be forgotten but it is not gone. DDT is out and the "Quads" are back. There are no African slaves or European explorers or settlers to import malaria into the New World, but there are immigrants from malaria-endemic countries entering in unprecedented number. So malaria emerges in little spurts of outbreak—a *frisson* from a former time. A group of Girl Scouts go camping in a California forest and get malaria. Mexican migrant workers in the area are suspected as the source. Soldiers who had never been outside the continental United States get malaria at Fort Campbell, Kentucky. Vietnam veterans are suspected as the source.

And in a real reversal of history malaria comes north—to the New York area! In 1993, two women and one man from the New York City borough of Queens are diagnosed as having malaria. All lived within two miles of each other (but were not acquainted). None had a history of travel that could account for the infection and it was decided they got it in Queens. Every immigrant in New York is suspected as the source. Prof. Curtis Patton, an epidemiologist from Yale University's School of Medicine is interviewed by the press and he says, "I think New Yorkers will be surprised because we like to think of parasites as tropical and malaria as tropical and it certainly is not."

Then there is the latest malaria news from New York; in the *New England Journal of Medicine* of July 7, 1994, there is an article entitled "Brief Report: Malaria Probably Locally Acquired in New Jersey." The authors report two cases of *Plasmodium vivax* malaria, one patient a 29-year-old woman, the other patient an 8-year-old boy, both from "Quad"-carrying counties of New Jersey. Again, immigrants are suspected, among them the "growing number of unskilled laborers, many of whom work in industries with no employer-provided health benefits" from Mexico, Honduras, India, West Africa, and so on.

Fear not; minioutbreaks are still curiosities. Malaria will not return to its former fearsome level. We are too rich a country to ever allow that to happen. So, for all practical purposes malaria in the United States is dead—but it ain't no Dodo.

Chapter 11

Merrie, Malarious Olde England: Before 1930

EVERY DAY WAS August Bank Holiday when the hippopotami frolicked in the Thames Estuary 100,000 years ago. England was not yet humanized; the first inhabitants—Bronze Age Celts or Picts—came from the European mainland at some uncertain date thought to be between 2000 and 1000 B.C. However, in the tropical prehuman England of the hippopotamus, the Anopheles vectors of malaria were breeding in the marshes and fens waiting the long years for the parasite. England has cooled, but those anophelines still breed in the Kent and Essex marshes and it is only in the last 50 or so years that there has finally been an end to indigenous malaria.

Malaria's arrival date and its human vehicle are unknown. Did the malaria parasite enter in the Picts or Celts? Malaria has been entrenched in northern Europe from ancient to modern times. The last homegrown German case occurred, in Schleswig-Holstein, in 1950. The Italians may have given malaria to England. Until Mussolini undertook his massive drainage-agricultural improvement project, the pontine marshes about Rome were, since antiquity, one of the most deadly malarious places in the world—tropical Africa notwithstanding. There were a lot of Italians in London and its environs from 55 B.C. to about 500 A.D. Then, there were reintroductions from returning soldiers and colonial officers. The malaria epidemic of 1917–1919 in Kent and Sussex stemmed from returning residents who had fought the Hun and the malaria parasite in Macedonia and Mesopotamia.

The Middle Ages were the Malarious Ages for England although it is difficult to make an allocation as to what proportion of the fevers and illnesses was due to malaria and what was caused by other infections such as influenza, typhoid, and typhus. Hippocrates and the physicians who followed him recognized the syndrome of periodic fever that attacked the victim every third or fourth day. The disease was called (until Horace Walpole wrote his tourist complaint in 1740, of "a horrid thing called mal'aria that comes to Rome every summer and kills one") intermittent fevers, remittent fevers, or commonly in England and her colonies the ague, a corruption of the Latin *febris acuta.* A Frenchman, struggling with the phonetic-spelling peculiarities of the English language, once remarked, "A plague on one-half of the English language, and an ague on the other half."

It is known that something had made the medieval Englishperson thin blooded and that the anemia-causing endemic benign tertian malaria would be a prime suspect. In 1973, a Yorkshire parish church, St. Helen-on-the-Walls, serving the faithful since about 1100 A.D., had to make way for a commercial-residential redevelopment project. The cemetery was dug up and over 500 skeletons from Middle Age burials exhumed—a treasure trove to the paleopathologist. Chronic anemia leaves its post mortem signature in the bone, a spongy, porous aberration known as porotic hyperostosis. When Dr. A. L. Grauer of Loyola University of Chicago's Department of Anthropology examined the skeletons, he found that 60 percent exhibited porotic hyperostosis. Grauer was uncertain as to the cause(s) but suggested that the abysmal poverty of that Yorkshire village and its handmaidens of squalor, near nonexistent sanitation, inadequate nutrition, and parasitic and microbial infections—to which I would add malaria—combined to make a short life that averaged about 40 years. It is precisely these factors that now hold the impoverished Third World of the second millennium in thralldom as they did St. Helens-on-the-Walls in 1100 A.D.

With the introduction of quinine in the mid-seventeenth century the delineation of fevers into malarial and nonmalarial types became clearer. There are drugs with narrow activity on specific diseases. This allows for a kind of diagnosis by therapy; if the drug is curative, then that patient must have had that disease. And so it is with quinine; it cures malarial fevers but not fevers of other origins.

The first European discovery of quinine in Peru or Ecuador remains shrouded in myth and mystery. Myth aside, in the 1630s it was brought to Rome where Popes and Cardinals were dying of malaria by Jesuit missionaries. It became known as Jesuit Powder or Cardinal's Powder—the drug of fashion for the ague-struck fashionable, dispensed by doctors, quacks, and doctor-quacks. It had been popularized in London and endorsed by the president of the College of Physicians, a Dr. Prujean, by 1658. In that year the Protector-dictator, Oliver Cromwell, lay acutely ill with the ague. He was offered the Jesuit Powder, but the intensely anti-Catholic Cromwell rejected anything tainted by that religion, even by a vernacular name of a drug. True to his Puritan God, the feverish, delirious Cromwell died.

Twenty years later Charles II didn't have to test his religious scruples when he had the ague. There was a Londoner, the nonphysician Robert Talbor, who was touting himself as the agent for a secret, miraculous ague cure. Over the objections of his doctors, Charles called for quack Talbor who did indeed provide the miraculous curative. The grateful Charles not only knighted Talbor but sent him to the French royal family whose numbers were being reduced by malaria and the succession threatened. Talbor administered his curative potion—Jesuit Powder or quinine disguised in wine—and so saved Louis XIV. The grateful Louis gave Sir Robert 3,000 gold crowns and a lifelong pension. Unfortunately Talbor didn't live long enough to enjoy much of the pension; he died at the age of 42. Self-promoting to the last, Talbor erected a prefunerary monument in Trinity Church, Cambridge, on which he had had inscribed, "The most honorable Robert Talbor, Knight and Singular Physician, unique in curing fevers of which he had delivered Charles II King of England, Louis XIV King of France, the most serene Dauphin princes, many a Duke and a larger number of lesser personages."

The epidemic-endemic malaria that gripped England's eastern coast from the time of Elizabeth I to that of Victoria was eloquently described by those two astute epidemiologists, William Shakespeare and Charles Dickens, who recognized the association of marshes and ague. *The Tempest*'s Caliban says, "All the infections that the sun sucks up From Bogs, fens, flats, on Prosper fall and make him By inch-meal a disease." Two and a half centuries later in *Great Expectations,* the concerned Pip says to the fleeing convict, "I think you have got the ague. You've been lying out on the marshes, and they're dreadful aguish." The ague must have been familiar to London's Boz. Between 1850 and 1860 there were more than 60,000 admissions for the ague to London's St. Thomas Hospital alone. About 1720 the quinine tree had been discovered growing on the eastern side of tropical South America. Quinine bark for processing now became more accessible and easier to transport to Europe. The price dropped and its use came to the commoner. Thus, those

60,000 patients at St. Thomas had been therapeutically diagnosed as having the real thing—the malaria ague.

The malaria of England is an enigma. For almost 100 years it has perplexed malariologists who still can't provide sure explanations for a number of the epidemiological phenomena.

(1) Why was malaria so spotty in distribution when the malaria-carrying anophelines were so ubiquitous? In 1929 the malariologist S. P. James drew a map of where the anopheline potential malaria transmitters were found and where, since 1860, malaria cases occurred. A great part of eastern England was mosquito ridden, but within that area, malaria was confined to a relatively few localities in rural, marshy Kent, Sussex, and, particularly, the wash of the Thames Estuary.

Explanations offered

Not all species or strains of anophelines are efficient malaria transmitters. In some, it is by fault of their biology; the parasite is loath to develop in them. In others, it is by fault of their inherent behavior, especially in that they prefer to partake of animal blood rather than human blood. Perhaps there were suitable species, subspecies, or variants of English anophelines in only a few, circumscribed locales.

Perhaps it was the human and not the mosquito that contributed the crucial factor. Malaria cases were not distributed randomly within the endemic areas but rather tended to occur in poor families living together. These were know as malarious houses. Until the 1920s there were impoverished marsh dwellers who lived in dark, dank, thatched, one-room hovels. This was a protective, nurturing environment for the mosquito—warm, damp, and her food supply all collected in one place. The mosquito could even domesticate over winter, still transmitting malaria when Christmas came.

(2) Why was the "benign" species of the malaria parasite, *Plasmodium vivax,* present in England so potentially lethal? It is generally agreed that the only species of indigenous malaria parasite in England was *Plasmodium vivax,* the cause of benign tertian malaria. Vivax malaria is responsible for high fever, anemia, prolonged lassitude but it is self-limiting—benign—it doesn't kill. And yet, even during the early years of this century, when malarial fevers could be reasonably well distinguished from other fevers such as influenza, the case fatality rate was about 5 percent.

English syphilitics also gave witness to *Plasmodium vivax*'s peculiar pathogenicity. In 1915, a time before there were antibiotics to treat late-stage syphilis, a Viennese physician, Julius Wagner von Jauregg, came up with the idea of "burning" out the treponeme from the central nervous system. His logic for this therapy was the experimental observation that *Treponema pallidum* died at a temperature that, while high, would not be lethal to its human host. The problem of how to elevate body temperature was solved by inoculating the patient with *Plasmodium vivax,* allowing several cycles of high fever, and then curing with quinine. Malariotherapy institutions for halting paresis's progress came into being in many parts of the world, including the institution at the Horton Mental Hospital, England. By 1925 over 10,000 English paretics had been given malariotherapy at Horton. It had been assumed that the procedure, although it was uncomfortable, was as benign as the parasite itself, and yet S. P. James, who had been the director of malaria research at Horton, cites a 1929 government report of a startlingly high treatment-related mortality: "Among patients suffering from general paralysis who undergo an attack of benign tertian malaria in the hope of curing their mental disease, the fatality attributable primarily to the malarial attack is between 10 and 12 per cent."

Explanations offered

Perhaps *Plasmodium vivax* is not as benign in nonimmunes as is conventionally thought.

Perhaps there was a peculiarly pathogenic strain in England.

There really is no satisfactory explanation.

(3) Why did malaria disappear without implementation of any special antimalaria programs? From 1860 malaria continually declined until it vanished from England sometime between 1930 and 1940. There was no grand scheme to control or eradicate it; it just disappeared and medical historians as well as malariologists have been asking why.

Explanations offered

Economic betterment acted as an effective antimalarial. When, in 1955, the World Health Organization began to consider embarking on its bold global malaria eradication program, it was as much a war between malariologists as against the parasite. The sociocrats held that good housing, good health and education services, and modern agricultural practices would make malaria disappear, as they had done in England and the United States. Against them were the technocrats, who won the day, proclaiming a world sprayed with DDT was the answer. When DDT failed, the chemotherapeutic troops were called on. The sociocrat-technocrat malaria wars continue today. However, there is no doubt that the rising standard of living after 1860 contributed significantly to the abatement of the English malaria problem. Better houses were built with more rooms that were better barriers to mosquito invasion. With more rooms, the family was dispersed to offer fewer targets for the mosquito.

By 1860 the price of quinine had fallen enough to be affordable by the rural poor. Chemotherapy at each attack finally killed off the parasites circulating in the population.

Finally, there is the turnip theory of malaria's eradication in England. The distinguished British medical entomologist of a generation ago R. A. Senior-White proposed this account for malaria's English demise. The vector, *Anopheles maculipennis,* prefers feeding on animals, particularly cattle; humans are its second choice. Before 1860 there wasn't enough fodder to feed the cattle herds over the winter and many of the animals were slaughtered before the cold weather set in. The overwintering mosquitoes, taking refuge in the house, had no choice but to feed on human blood. Around 1860, the turnip was introduced as an easy crop to feed domestic animals. With the turnip the cattle could be maintained in the cowsheds, usually adjoining the human habitation, over the winter. In these warm, dank sites filled with the blood supply they craved, the mosquitoes would gather and thrive; the humans would be spared and malaria transmission would be halted.

From the United States, across the Atlantic longitudes to London, the most tropical of tropical diseases, malaria, continued endemic into this century. In 1865, an outbreak of that other most tropical of tropical diseases, yellow fever, revealed how porous Britain was to invasion by foreign microbes.

In the rapidly emerging nineteenth century Industrial Age, Swansea in Wales and Santiago in Cuba were economically tied sister cities. Swansea, an ocean port near the Welsh coal fields, developed a prosperous iron smelting industry.

Cuba held a cheap, abundant supply of iron ore, and a lively maritime trade arose—Cuban ore to Swansea, British manufactured goods to Cuba.

On July 28, 1865, the *Hecla,* a modest wooden cargo ship laden with iron ore sails from Santiago, destined for Swansea. For the *Hecla* this is an ill-fated journey from the moment it raises sail in Santiago. Two of its sailors have already died of yellow fever while in port. During the trip three more sailors sicken. They die and are buried at sea.

On September 8 the *Hecla* arrives in Swansea. On board is the desperately ill sailor, James Saunders. A physician is called, but the ship's captain disguises the situation by telling him that Saunders suffers from dropsy. Nor does he disclose the earlier deaths. The doctor, a shrewd port physician, wise in the ways of ships' captains and owners, diagnoses Saunders as a case of yellow fever. Quarantine of the *Hecla* is recommended. Saunders dies and is quickly put under earth.

By 1865, Britain had already paid a medical price for its lively overseas trade. There had been three outbreaks of Asiatic cholera, and in response, quarantine laws had been enacted. However, Parliament was influenced by powerful shipping interests, and these laws and their enforcement were weak. The *Hecla* was not ordered out of Swansea; the yellow jack did not fly from its mast.

On September 15, a week after the death of Saunders, a dock worker who has been unloading the *Hecla* sickens. His home is some distance from Swansea itself. The worker dies of yellow fever. A week later, September 22, the first Swansea resident dies of yellow fever. The person has no connection to the *Hecla* or the dock area. More Swansea citizens contract yellow fever. By October 5, 29 people have contracted the infection and 16 have died of it. The cold of fall sets in; the outbreak ceases and no more cases occur in Swansea.

Saunders must have become infected at sea during the latter part of the 39-day voyage home since yellow fever's incubation period—the time from acquiring the pathogen to the onset of symptoms—is 3 to 10 days. He would have been bitten by one or more virus-carrying *Aedes aegypti* mosquitoes living in *Hecla*'s humid holds and breeding in her exposed water storage barrels. The first Swansea case, the dock worker, must also have become infected from the *Hecla*'s mosquitoes. How the 29 citizens of Swansea got their yellow fever is open to conjecture. Did the mosquitoes disperse from the *Hecla*? Did the *Aedes* become established in breeding sites in and about Swansea until the cold weather eradicated them and the yellow fever virus they harbored?

Plague, cholera, malaria, and one outbreak of yellow fever were Britain's own infectious diseases with a somewhat tropical tinge. Britain has also had almost four hundred years of self-interest in the tropics. The first book, more of a pamphlet really, was published during the reign of Elizabeth I, *The Cures of the Diseased in Remote Regions, Preventing Mortalitie, incident in Forraine Attempts of the English Nation,* by Humphrey Lownes (1598). During all that time of *forraine attempts,* there have been outstanding men and women serving overseas and at home, laboring in the field and laboratory to make the torrid zone a healthier place. Tribute should be paid to those servants of science, especially so in 1998, the centenary year of Ronald Ross's discovery, in his modest Calcutta laboratory, of the transmission of malaria through the mosquito. That supersensitive, single-minded man went to his grave still holding the conviction that malaria and other vector-borne scourges could be eradicated if only the weak-willed governments would commit to exploit the gift of his and other scientific discoveries. In this sentiment, Ross lives on.

Chapter 12

Paradise Lost 1945 to 1996

WE'RE ALL RIGHT. The big three tropical diseases have now departed although the occasional case of hookworm and malaria remind us, like faded souvenirs in a collectibles shop, of their former endemic time. Our good neighbors to the south are not so all right; malaria worse than ever since it is now multidrug resistant, hookworm as prevalent as ever, and jungle yellow fever remain as old familiars. But we're all right. Except for a few new nasties.

Had they not appeared in respected medical research journals, their titles could serve as headlines of sensational articles in the *National Enquirer* genre:

"Babesiosis in Washington State: A New Species of Babesia"
"A Suburban Focus of Endemic Typhus in Los Angeles: Association with Seropositive Cats and Opossums"
"The Malaria of Airports" (this was Paris but it could equally apply to the United States)
"Epidemic Cholera in the Americas"
"AIDS: An Old Disease from Africa"
"Leishmaniasis in Americans"
"American Trypanosomiasis (Chagas' Disease)—A Tropical Disease Now in the United States"
"Neurocysticercosis in an Orthodox Jewish Community in New York City"

That last title, a paper by Peter M. Schantz of the Centers for Disease Control in Atlanta and his colleagues, especially caught my attention. In 1980 I wrote an article for *Natural History* magazine, "On New Guinea Tapeworms and Jewish Grandmothers" (it became a chapter and the title of a book published by W. W. Norton in 1981). It was a story of a transcultural tragedy whereby a primitive Irian Jaya (Indonesian New Guinea) pig-dependent highland tribe came to be infected with the larval stage of the pig tapeworm, *Taenia solium.* The parasite was first introduced as infections in the Balinese soldiers sent to persuade the tribe, the Ekari, that subjugation to the Java government was good and necessary. Later, the infected Balinese pigs, sent as a gift to soften the military persuasion, added to the source of worms.

The Ekari nearly became the tribe that lost its head. The larval tapeworms went to the brain (a condition called *neurocysticercosis*) and caused epileptic-like fits. The seizures often occurred at night as victims slept next to the fire; its very cold at night in the Irian Jaya highlands. There were horrible burns. Deaths. It was a worm-driven disaster.

I wanted to show readers that the transcultural infectious transaction that befell the Ekari was not unique to politically weak, naive primitives; rather, we are all at risk to the phenomenon, even Jewish grandmothers in New York City. Unwittingly, these ladies acquired a 45-foot-long intestinal tapeworm, *Diphyllobothrium latum,* in the course of making *gefilte* fish. The tapeworm was originally endemic in the Baltic and Scandinavian countries where it was passed from person to person through the larval form in freshwater fish. Immigrants from Scandinavia settled, together with their tapeworms, in the Minnesota-Michigan lake regions where they carried on their traditional occupation of fishing. The fish in those lakes became infected, and when shipped live to New York and other East Coast cities, they then infected the Jewish grandmothers who while making *gefilte* fish, tasted some of the preparation while it was still quite raw (and the tapeworm larvae still quite alive) to determine its doneness and in so doing got that 45-foot sucker.

A rare case of *Diphyllobothrium* occurs now and then, but the gefilte-fish-making Jewish grandmother is a fast-disappearing breed of woman. Anyway, the stuff comes in cans and bottles now and is on the supermarket shelf. Nor was there any doctrinal conflict in having a *Diphyllobothrium.* It was, presumably, a kosher parasite that came from a kosher fish (I'm probably over my head on this one and should seek rabbinical advice). The Schantz report was a worm of quite a different theological import. It was a *pig* tapeworm. And what's more it was in members of the Jewish ultraorthodox Lubavitcher community.

Between 1989 and 1991 four frightened people—a child of 6 years, a girl of 16 years, and a male and a female in their thirties—were brought to the emergency rooms of New York City hospitals. Each had had a seizure, convulsions that left them unconscious. None were epileptics; none had ever previously had a seizure. In one case, the 16-year-old girl was struck dumb, aphasia, before the seizure. For the others, the convulsions were of sudden onset with no preliminary symptoms. All were Orthodox Jews.

Each person was referred to the hospital's neurological service workup and each put through the diagnostic wringer. First the history: No, none had a history of epilepsy. Yes, this was the first seizure each ever experienced. No, none had received a blow to the head or suffered from a febrile illness prior to the seizure. No, none had ever traveled to some far-off exotic or tropical country. They were all nice, normal Lubavitcher Gothamites, right down to the man's big black beard. This left the brain tumor as the most logical explanation. Each patient's head sat for its medical portrait by CAT scan and magnetic resonance imaging to confirm and locate the tumor. In each case, round lesions, 7 millimeters to 2 centimeters in diameter, were seen in the brain. The opaquely dense lesions appeared calcified.

This wasn't the image of a tumor. Neurological tests failed to indicate a cancerous cause of the convulsions. Obviously, those round, opaque things in the brain were responsible. But what were they? Very few conditions give rise to such globular, foreign bodies in the cerebrum; two of the most notable are parasitic worms. One, *Paragonimus westermani,* which humans get by eating raw or marinated (but uncooked) Asian freshwater crabs, was an unlikely etiological candidate. These people had certainly not been eating crabs of any kind, Asian or American, cooked or uncooked, since *all* crabs were *trafe*—forbidden by dietary law.

The other parasite was an even more unlikely possibility, being a tapeworm parasite, *Taenia solium,* that came from the super*trafe* pig. And yet, the scan images coincided with textbook descriptions of neurocysticercosis, the infection of the central nervous system with the larval, cystic form of *Taenia solium.* Just like the ill-fated Ekari of New Guinea. The still-doubting neurologists sent serum samples to the Centers for Disease Control (CDC) for serological testing. The immuno-parasitologists at the CDC detected antibodies specific for *Taenia solium* in the patients' serum samples; those people were infected with the parasite and their immune systems had responded by elaborating the specific, signature antibodies. Now the astounded, but not-so-doubting neurologists had the surgeons open the heads of two patients and pluck out a cyst from each patient's brain. Under the pathologist's microscope sat the unmistakable bladder-like larva of

Taenia solium. It was an epidemiological mystery. How could these observant Brooklyn Jews ever become infected with the larval stage of the pig tapeworm?

If I were a parasitologist fated to return to life as a parasite, I wouldn't want to come back as a pig tapeworm—such a hazardous existence, perpetuation depending on the several creatures to be my host and their unsanitary toilet and eating habits. In the *normal* course of its life cycle, the adult *Taenia solium* exists as a 6- to 20-foot-long segmented "tape" attached by its sucker and hook-adorned head to the wall of the human small intestine. Only a human will do for the lengthy adult worm; it is a very host-specific stage.

The segments at the end of the tape are filled with eggs. These gravid segments break off, disrupt to release the eggs, and are passed in the feces. Not everyone in this uneven world has a flush toilet; they defecate where they can. If feces were fluorescent, the whole world would glow at night. Also, farmers of the Third World will use feces, euphemistically called night soil, as fertilizer. Pigs just love feces, and if they dine on some containing the *Taenia solium* eggs, the eggs hatch in the intestine and release the larvae which migrate through the body and usually reach the musculature (but they may be filtered out in many parts of the pig's body—brain, eyes, and so on) where they round up and grow in the cysticercus stage. There they remain quiescent until the pig is eaten in the form of sausage or undercooked pork by the human. In the human intestine, the cysticercus head pops out (figure 3). From this germinal head (the scolex) the line of segments are formed one on another.

That is the natural path of *Taenia solium*'s life cycle—from human to pig to human. However, of all the tapeworms infecting humans, *Taenia solium* is unique in that the human can be a stand-in for the pig; that is, the worm can exist in

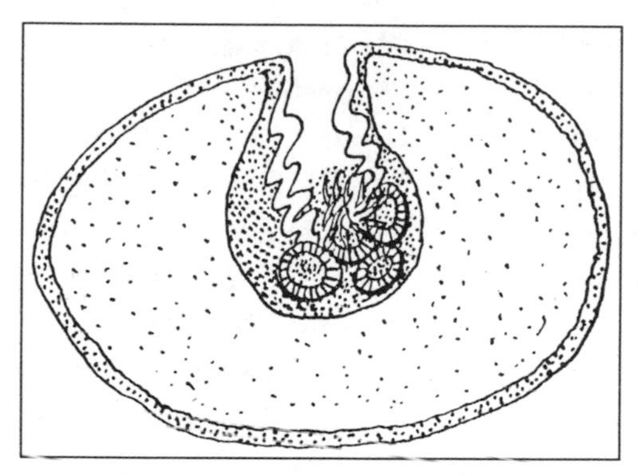

FIGURE 3
The pig tapeworm cysticercus

humans in both the adult tape and larval cysticercus forms. For example, if a human accidentally swallowed the egg of the closely related beef tapeworm, *Taenia saginata* (the adult "tape" is in humans and the larval cysticercus is in bovines), nothing would happen; the egg or newly hatched larvae would die. However, if that human ingested an egg of *Taenia solium,* it would hatch, the larvae would survive, migrate in the body, and develop in the same manner as if that human were a pig. It follows that if a person has cysticercosis, he or she must have ingested feces from a person intestinally infected with an adult, "tape"-stage *Taenia solium.* This can unwittingly occur when eating a feces-fertilized salad green. It more often occurs in a hand-to-mouth transfer from fecally contaminated hands. Fecally contaminated hands frequently belong to people with primitive toilet habits. This is not (God forbid!) the toilet habits of the Brooklyn Lubavitchers, and the mystery of the minioutbreak of neurocysticercosis among them deepened.

Peter Schanz and his epidemiologists took up the case where the neurologists and immuno-parasitologists left off.

The main tool of the epidemiologist is his or her inquiring mind. An epidemiologist probes, asks questions, noses around, and then tries to put two (or whatever number it takes) and two together. The CDC epidemiologists may have come to their inquiry with an image of the Lubavitchers as an unworldly, intensely Talmudic-scholarly group living in isolated, genteel poverty in some Brooklyn ghetto. This preconception was only partially true. The Lubavitchers are scholars of the Talmud and they do follow the tenets of their religion to the last letter, but unworldly, isolated, and impoverished they are not. Many are successful businessmen. They carry beepers so that when their Messiah arrives on Earth, they can immediately communicate the event to one another. These are middle- and upper-middle-class people. And like many of the middle class, they had housemaids. Mexican housemaids.

Mexico is the world's pig tapeworm capital, possibly excepting Bali. Almost 2 percent of all deaths in Mexico have been due to neurocysticercosis, and it has been estimated that about 4 percent of all Mexicans have the adult *Taenia solium* in their intestine. There have been several drugs to expel the tapeworm, but until recently there has been no effective treatment for cysticercosis. Two relatively new drugs, praziquantel and albendazole, have proven to be very effective once the infection is diagnosed. That, of course is the catch—getting the Mexican patient who often lives in a remote village to a facility, such as the Neurological Institute in Mexico City, where expert diagnosis and treatment is available. The Mexicans speak of mounting an antitapeworm campaign, but that would require an extensive case detection and follow-up treatment system beyond their present capability. For long-term effectiveness they would have to implement a rigid inspection system of pork with all slaughtering done at government-monitored abattoirs. This is the requirement for meat processing in the United States where in 1990, for example, 83 million hogs were slaughtered and only 3 were found by the inspectors to harbor the larval *Taenia solium*, the cysticercus, in their flesh.

With those parasitological statistics of Mexico's tapeworm problem in mind, the CDC workers inquired whether any Brooklyn patients or their families had employed Mexican housekeepers or cooks. *All of them had Mexican maids.* In fact, later inquiry revealed that of the approximately 8,000 families making up the Brooklyn Lubuvitcher community, an astounding 94 percent employed immigrant, mostly Mexican, housekeepers.

Now to close the epidemiological circle. The CDC set out to find the Mexican maids and determine by direct parasitological examination of stool samples or indirect serological evidence of specific serum antibodies whether any of them had or had had taeniasis, an infection of *Taenia solium*. This was no easy sleuthing. Some of the patients' families had employed multiple maids, in one case ten different women housekeepers over the two years before the patient became ill with neurocysticercosis. Some former maids had returned to Mexico; one was tracked down in her home village and was found to be serologically positive. Others wouldn't give either blood or feces for testing. One Mexican maid, with a patient's family for ten years, had returned to her village in Oaxaca for vacation each year. She had taenia eggs in her stool specimen. When told of the finding, she fled. Too bad, she could easily have been treated and good help is not easy to find these days. Another serologically positive maid also fled her employ when advised of the results.

It was a small outbreak, but it was another signal to alert the medical community to look for unthought-of diagnoses when the signs and symptoms point to a seemingly un-American infection. As for the Lubavitchers, I wonder what their Talmudic scholars made or will make of all this. Who sinned? The pig? The Mexican maid? The parasite? The patient (even though an innocent host)?

Neurocysticercosis in any religious denomination will never be as American as the heart attack. There is, however, another infection that not only has the potential to become a serious health problem; it would, in fact, *be* the heart attack. Chagas' disease, caused by *Trypanosoma cruzi* as noted earlier, is a major cause of heart disease in the vast endemic belt of Latin America from Argentina to Mexico. You may also recall that this parasite is transmitted by the blood-sucking triatomid bug. About 90 species of triatomid exist in the Americas and 30 of the species have been incriminated as vectors. Of these 30, 9 are in the United States. And those 9 carry *Trypanosoma cruzi.* The parasite is with us.

Scientists, since early in this century, knew that the bugs were in the United States and that there were *Trypanosoma cruzi*-like protozoa within the bugs' guts. They also knew that there were *Trypanosoma cruzi*-like organisms in the blood of many wild animals from Texas to Maryland to California. But was it the real thing? The real McCoy? Were these protozoa in the bugs and beasts *Trypanosoma cruzi* or were they some protozoan imposter of no potential threat to humans? Today, elegant noninvasive techniques separate and identify the look-alikes, the parasitic wanabees. However, in the 1940s when the early research on infections in the U.S. triatomids and wild animals were being undertaken, those methods were in the future and the best proof would be to inoculate with organisms taken from bug or beast to see whether it would take and, if so, whether typical Chagas' disease would follow. Then, as now, there was no effective curative drug for Chagas' disease and certainly no informed volunteer would wittingly agree to become a guinea pig. In retrospect, the 1940 account of an experimental infection in a human might be expected to have been published in a medical journal of Nazi Germany. It is something of a surprise to see the article, "Infectivity of the Texas Strain of Trypanosoma cruzi to Man" in the austere, careful *American Journal of Tropical Medicine.*

Dr. Ardzoony Packchanian was a medical parasitologist at the University of Texas's School of Medicine in Galveston. During the 1940s he carried out studies on the biology and cultivation of *Trypanosoma cruzi* and other related hemoflagellate protozoa. On December 5, 1940, he caught a *Triatoma heidemanni* bug in the Texas town of Three Rivers. Under the microscope the dissected bug's gut revealed a swarm of active *Trypanosoma cruzi*-like organisms. From the crushed bug a "part of the material was introduced into the left eye of a male negro," and also inoculated into three mice and two guinea pigs. The mice and guinea pigs developed a typical acute Chagasic infection. Parasites multiplied in their damaged hearts.

The African American also became infected. He became feverish. His lymph glands enlarged, and he developed the characteristic painful swelling of the eye, Romana's sign, in which the bug material was introduced. Packchanian in his published paper refers to this man as the "patient," but we are never told who he is other than "the negro patient at the time of exposure, was 24 years of age, weighed 133 pounds, and was 6′½″ in height" and was in "good physical condition" with no signs of cardiac pathology. But who was he? A convict prisoner? A paid dupe? Or even, most unlikely, a volunteer? The paper shows a photograph of a clean-shaven African American with an eye where "both upper and lower eyelids were swollen and hyperemic extending to the orbital margins."

The "patient" is followed for two years, declared healthy, and dropped from further study and care. Ardzroony Packchanian acknowledges his indebtedness to the director of the State of Texas Laboratory and the Drs. C. H. Standifer, G. G. Zeller, and J. Vanden Bossche for their clinical assistance, but there is not even the courtesy of a thank you to the "patient." Packchanian concludes that "the present study demonstrates clearly that the Texas strain of *T. cruzi* is infective to man and that the disease produced in man with this strain is clinically identical with that found in South America." In making that proof, Packchanian walked on the dark side of biomedical research as did the editors of the *American Journal of Tropical Medicine* in its publication.

The status of Chagas' disease today is unknown. Epidemiological exploration is not the kind of research now favored or funded. Infectious disease epidemiology has fallen on especially hard times, a poor competitive rival to molecular enterprises. Nevertheless, we know that all the Chagas' elements—the bug, the parasite, and its animal reservoir hosts—are in place throughout great areas of the rural, suburban, and probably, urban United States. Three cases of acute Chagas' disease contracted have been recorded within our national boundaries, two in Texas infants. The third case was not on the Mexican doorstep, but 120 miles east of San Francisco.

Since 1916 it has been known that *Trypanosoma cruzi*–infected triatomid bugs and *Trypanosoma cruzi*–infected animals—dogs, wood rats, ground squirrels—were common in California. However, the infection stayed in its animal orbit and was not considered a threat to humans. No cases of acute disease of chronic Chagasic heart disease had been diagnosed. Then in the summer of 1982 a 56-year-old woman from Lake Don Pedro, in east central California, went to her local physician. She wasn't an unusual patient—a typical "normal" sick person sitting in the doctor's waiting room. You can imagine the doctor telling her, "There's a lot of this illness going around now" and maybe thinking, "There's a lot of 50+-year-old women who feel fatigued, have no appetite, and may even feel feverish."

"Very sick," "normal sick," and even "imagined sick"—all patients are screened through the laboratory testing process. It helps in diagnosis; it is a defense against malpractice litigations and adds some income to the doctor who runs a lab-in-office. As everybody knows, to visit the clinical lab is to be stuck by a needle and blood taken. On a stained preparation of this woman's blood, the technician saw under the microscope something that shouldn't be there. It was a fishlike organism, about $1\frac{1}{2}$ times longer than the diameter of a red blood cell. It had a fine "thread" at one end and a prominent, dark staining, pop-eyed granule at the other end (figure 1B). The doctor was called, and remembering a lecture from parasitology past, he recognized the artifact as the blood-stage form of *Trypanosoma cruzi.* This was soon confirmed by the CDC's Division of Parasitic Diseases.

This was precisely the kind of case that the U.S. public health home guard, the Centers for Disease Control, investigates and, when necessary, takes appropriate action against. A U.S. woman who never used intravenous drugs, never had a blood transfusion, and never traveled to a Chagas'-endemic country south of the border had a parasitologically confirmed, clinically manifest infection of *Trypanosoma cruzi.* A human case of a new and rare (for the United States) infection makes alarm bells go off at the Centers for Disease Control, and it rapidly deploys specialist investigational officers.

Before being treated with the best of the inadequate drugs, blood was taken from the woman and the trypanosome isolated in a culture medium. The rapidly multiplying organisms were sent to Dr. Michael Miles at the London School of Tropical Medicine and Hygiene, an expert in identifying the strains and characteristics of *Trypanosoma cruzi* on the basis of their enzyme constitution. Miles identified the isolate from the Lake Don Pedro woman as a South American type (a zymodeme) acquired by humans, via the bug, from wild and domestic animal reservoirs.

With this clue from the parasite's enzymes, the CDC investigators rounded up the usual zoonotic suspects and bled them. From the blood of 2 of the 19 ground squirrels (*Spermophilus beecheyi*) *Trypanosoma cruzi* parasites were isolated in cultures. Dogs from the local pound produced no parasites in cultures, but 6 of the 10 dogs examined had specific antibodies. They were or had been infected.[32] From the nearby Merced County animal shelter, an astounding 24 of 28 dogs were serologically positive. Even in San Francisco, serologically positive dogs (4 of 27) were present. Obviously, dogs play a role in the epidemiology of *Trypanosoma cruzi* in the United States although the exact nature awaits discovery.

Serology is also an important companion to clinical epidemiology in human diseases. One case does not an epidemic make. But were any more out there? And what is our level of exposure. Blood samples were taken from 241 residents of Lake Don Pedro and 19, 7.9 percent, were positive for *Trypanosoma cruzi* antibodies. In the adjacent county, 4 percent of the residents' blood samples were positive as were 4 percent of the blood samples from San Francisco. None of

32. Failure to find the blood-stage form of the parasite by either direct microscopical examination or culture doesn't necessarily mean that the person or animal is not infected. *Trypanosoma cruzi* tends, after a while, to disappear from the blood and persist as a purely intracellular parasite. Hence the need for adjunctive serological diagnosis. For the infectious disease epidemiologist, if it moves, bleed it!

these positives had any signs of clinical disease yet. A disturbing result from an earlier investigation in rural Georgia had divulged that 7 percent of patients with myocarditis of unknown cause were *Trypanosoma cruzi* seropositive. In contrast, 0.4 percent of people in the same geographical area but without heart disease were positive. There seems to be a considerable number of native-born United States citizens who have had an encounter with the triatomid bug and the parasite it carries.

Is Chagas' disease a coming plague in the United States? Probably not. Could it be? Possibly yes. People are escaping from the cities to the bosky countryside—to its semidomesticated wild animals, bugs, and the bug's parasites. And in a reverse migration of the invertebrates, the bugs are coming to town. So far the strains isolated seem not to be virulent, but they can hot-up. And in the long run the avirulent types may not be so innocent.

New immigrants from hot-strain endemic areas have flooded into the United States and this worries Dr. Louis V. Kirchhoff, an established investigator of the American Heart Association at the University of Iowa's Department of Internal Medicine. Kirchhoff took blood samples from 205 immigrants from El Salvador and Nicaragua living in Washington D.C. Five percent were found to be infected with *Trypanosoma cruzi*—not the indirect evidence of serology but actual infection—parasites were in their blood. Kirchhoff extrapolates from these data and writes, "If this reflects the prevalence of infection among the general population of immigrants from countries with endemic disease, there may now be 50,000 to 100,000 immigrants with *T. cruzi* infection living in the United States." Many of these infections would be of virulent strains, which would attack the North American heart with South American aggressiveness.

So, here we have two "immigrant diseases." One, neurocysticercosis, is a curiosity (except for those infected), an example of how two peoples of widely diverse cultures, Orthodox Brooklyn Jews and Mexican Roman Catholic immigrants, can interact out of economic necessity and transact their infection in so doing. The other, Chagas' disease, is different. It is no curiosity but rather a continuing specter, a virtually untreatable disease that because it has fallen out of epidemiological fashion, is not being adequately monitored. Curiosity and specter—both reflect the conflicting U.S. attitudes toward immigration and immigrants—hospitality and xenophobia. Immigrants have made the United States great. Immigrants can be a nest of dangerous, menacing germs. Take your choice.

Chapter 13

The Twenty-first Century

LET US END as we began, worrying about the weather. You who shoveled your way through the blizzards of 1995–1996 may be reluctant to believe they were manifestations of global warming. But the meteorologists assert the warming atmosphere causes more evaporation from the oceans and other great bodies of water; this condenses to rain, sleet, and snow. Despite the snow and frigid cold, the world is warming. The Big Thermostat is out of whack. There will be more cataclysmic hot and cold storms, more hot summers, more drought, more floods, and more disease.

Meteorologists, epidemiologists, sociologists, agronomists, and economists all predict a bleak second-millennium world. The right people are now voicing the right sentiments. Temperature-raising atmospheric pollution must stop. Deforestation and ecological degradation must stop. Population growth must stop. Poverty and its yokemate, ill health, must stop. But despite logic, humankind has yet to kick its bad old habits. The projections for the twenty-first century are forbidding. We will keep breeding like humans; the world population will increase to 9 billion by 2025. There will be continued if not more political turmoil and terrorism. The civil unrest will cause more desperate people to flee to the United States and other countries of the industrialized west. In 1978 there were 5 million political refugees; in 1992 there were 17 million. Double that number for economic refugees.

Make a graph, draw a projection line from those figures to 2100 A.D., and draw your own conclusions.

The carbon dioxide level in the atmosphere will double by 2100 A.D., and its greenhouse effect will elevate global temperature by about 3 to 4°C. Ice will melt, seas will rise, and if all goes according to schedule, maybe 15 to 20 percent of the coastal, arable, habitable land will be inundated. It will be a wonderful world for the insects and the diseases they carry.

I remember reading, many years ago, Byron's poem "Don Juan." For some reason the only line that was fixed in my memory (and probably not with faithful accuracy) is, "What gods call gallantry and men adultery is much more frequent where the climate's sultry." Perhaps the cause for this poetic memory is less Freudian than professionally biological. The phenomenon Byron alludes to can be expanded to include a panoply of creatures, with and without backbones, who become sexually aroused, more reproductive, when the temperature rises. This is of concern to those whose occupation, like mine, is concerned with insect-transmitted infectious diseases. The more insects, the more disease(s).

When the British medical entomologist J. D. Gillett, a mosquito expert, wanted to make the point of the interconnectedness of reproduction rates and temperature, he took for his example not a blood-thirsty anopheline but the harmless lepidopteran with the threatening name, the Death's Head moth. The Death's Head moth has a geographical range that extends from the temperate zone to the tropics. In the temperate habitat it takes the moth six to eight months to "turn over," to go from egg to caterpillar to pupa in the cocoon and finally to adult moth. In Africa that cycle takes 21 days. Moths or mosquitoes, heat makes them breed faster. The warmed-up twenty-first century will have not only a greater human population but it also will have an infinitely greater population of insects of medical importance. They will intrude farther and farther into northern and southern temperate zones. The malaria-transmitting, temperature-dependent Anopheles mosquitoes, which require a minimum of 16°C to live and breed, will create malarious Maine and Siberia. There will be fewer cool nights to chill the biting insect's thirst for blood. There will be less respite, day or night, month to month, from the attack of the hematophagous insects.

In the warm, humid greenhouse climate of the future, the mosquitoes et al. will also live longer and bite more frequently during their extended lifetime. This increases the transmission of the diseases they carry. Moreover, the viruses and parasites that cycle through these insects will develop and multiply faster themselves within their warmed-up invertebrate hosts; this will make them still more ready to be transmitted to the human. And when they are, it will be with a greater infecting dose and consequent greater disease severity. Infectious disease experts agree on the likelihood that arboviruses, notably dengue (breakbone fever) will come north. Other experts foresee new outbreaks of yellow fever in the United States.

The infectious disease oracles who take this apocalyptic view of futureworld predict that there will be more and more "God-knows-where-they-come-from" diseases. In the last half of this century, humans have been beset by new microbial and parasitic infections. Not virulent varieties of old familiars (although this too has occurred) but seemingly new pathogens—Lassa fever, Ebola fever, Lyme disease, Legionnaire's disease, Hantavirus disease, Erlichiosis, Babesiosis, and, paramountly, AIDS. Are they but beginning points on our time graph? Will the twenty-first century be a time of epidemics of strange and terrible new diseases? And if so, where will these new-to-humans pathogens come from? What circumstances will give them birth? Where will they be born? In a forest in India, a twentieth century outbreak occurred that might well be the harbinger of what could happen in an ecologically and socially deranged twenty-first century.

A pristine forest supporting a rich population of wildlife, the Kyanasur forest had once covered the low hills of India's Karnataka state. Large troops of languor and macaque monkeys scurried through the trees. Little human intrusion had occurred during the colonial and long precolonial eras other than a few small villages at the forest's edge. Independence seemed to confer the license for riotous reproduction. Population grew. More and more villages now surrounded the Kyanasur forest. Some settlements invaded the forest itself. Fuel was needed to supplement the customary cow dung. Villagers went into the forest; trees were cut. In the thinned woods a thick undergrowth of Lantana bush pioneered into the clearings. More and more cattle were sent into the degraded forest to forage.

This was the condition of the Kyanasur forest in March 1957 when the villagers who went into the forest to collect firewood found a great number of monkeys lying dead on the forest floor. That same month people of the forest villages began to sicken and die. The illness opened with a high fever followed by severe back and joint pains, an irregular heart beat, diarrhea, vomiting, and bleeding from the orifices. Initially, about 20 percent of the ill died. Later, with better management the case fatality rate was reduced to 5 percent. Those who recovered were weak and dispirited for months.

When monkeys die and people die, the first suspicion is of yellow fever. Yellow fever has been a specter that has long loomed over Asia. It has never come to that continent. But that doesn't mean it never could; the potential vector mosquitoes are there. The investigators immediately noticed that the dead and dying monkeys were heavily infested with ticks. From the blood of the monkeys and sick villagers a virus was isolated. It was related to yellow fever but it was not yellow fever. It was a virus completely new to the virologists—a tick-transmitted arbovirus of the tropics. Nothing like it had been seen before. Armed with this knowledge, epidemiologists now began to unravel the factors of the new disease they named Kyanasur forest fever. The scenario of Kyanasur forest fever involved a procession of animals, humans, and ticks, set in a shifting ecological tableau. The story line went like this:

The cattle came to the forest carrying the adult ticks with them. The female ticks, engorged with blood, dropped to the ground and laid their eggs. The eggs hatched to become larval "seed ticks." The hungry baby ticks began their own search for a blood meal, usually that of a forest rodent. In former times this was a small-time affair; there were few cattle and few rodents in the forest and, thus, relatively few ticks. Then the biological balance tilted. There were more people in and around the forest. There were more cattle to feed on less and less available arable land. The cattle-owning families sent their animals into the forest to forage. The forest was degraded and this caused a pivotal ecological-epidemiological change—the intrusive growth of Lantana thickets. These thickets were an ideal habitat for rodents and they bred prodigously. And so, of course, did the ticks.

Somewhere in this mélange was the tick-transmitted virus. The original host is unknown. It may have been in rodents or even in birds on which the ticks also fed. The virus was benignly in balance with these animals as it was with cattle. The cow became a dangerous amplifying host. Monkeys and humans were creatures of a different susceptibility. The monkeys descended from the trees to disport and feed on the forest floor. When they did so, they became infested with the Kyanasur forest fever virus–infected ticks. In the monkeys, the virus expressed virulence. The monkeys died. Ticks attacked the other primates, the humans, when they came foraging in the forest. The humans also sickened, and many died.

And that is how a new disease is born. The birth is brought about by human-made habitat changes that alter behaviors and populations of animals, pathogens, and transmitting arthropods. If these ecological-epidemiological derangements continue unmodified, then we may be in for a bad time in the twenty-first century. Kyanasur forest fever; Legionnaires' disease; Lyme disease; Ebola, Lassa, Marburg, and Hanta viruses; and AIDS may only be curtain raisers for the main act.

If we accept that the threat exists, then should we begin to provide the resources to meet the coming threat? And even if the money and will are there, will science save us? Indeed, *can* science save us? There are problems, especially with the customary first line of defense, the quick drug fix.

The infectious disease therapy projection line on our time graph is now flat or falling. Although the panic button has not yet been hit, many of the old standbys, the tried and true chemotherapies for microbial and parasitic diseases, are failing. The common bacteria, the staphs and streps, are becoming more and more resistant to more and more of the formerly effective antibiotics. The antibiotic nova of the 1950 to 1970 decades is now a dying dwarf star. Fewer and fewer alternative therapies are being discovered. Unless there is a revolutionary breakthrough like that which ushered in the antibiotic age, the twenty-first-century doctors may be as helpless as those of the nineteenth century were in dealing with pneumonias, strep throats, diarrheas, and cuts and bruises. It is also worth noting that in the twenty-first century tourists to the tropics might acquire malaria for which there will be no cure.

The search for new antimicrobial and antiparasitic agents is so tediously difficult and so unglamorous that it has not been pursued with the resources of expertise and funding that would seem prudent. Academic and governmental researchers (aside from the military who *know* what is needed to combat their microbial enemies) are not particularly interested in chemotherapeutic research. It is so unfashionable and, worse still, so ungrantable. The pharmaceutical industry is somewhat more active, but infectious disease chemotherapy takes second place to the search for new good drugs for depression, heartburn, high cholesterol, and heart attack. These are profitable sources of morbidity. The unprofitable infectious diseases are orphan diseases. Malaria which kills 2 to 3 million each year is an orphan disease. Ebola which kills almost all it strikes—in Africa—is an orphan disease. The fear of future epidemics of new and terrible diseases lurks in the public mind, but the public, government, and even the pharmaceutical industry trusts that science would work its expected miracle, like the 24-hour discovery of the cure that saved the United States in the movie *Outbreak*. As my English colleagues would say to this notion, "The best of British luck to you." Science, biomedical science, simply does not function in so efficient and focused fashion.

Every biomedical scientist knows that the great, beneficial discoveries come *both* from logical, rational, dogged investigation *and* from serendipitous, empirical, seemingly remote, unconnected findings. More often than not, logic follows serendipity. We now know that the pivitol cell in immunity is the lymphocyte. Lymphocytes do almost everything; from them come antibodies as well as the chemical signals to turn the off-and-on switches of other cells of the immune system. A subpopulation of lymphocytes, given the right signal, can develop into cancer cell killers. And yet, 50 years ago the lymphocyte was a mysterious nobody in the blood. My old college hematology book suggested that it might transport fat. In 1956, a poultry science graduate student was doing his dissertation research on the bursa of Fabricius, a bit of lymphoid tissue in the chicken's rectum that few others had studied and whose function was unknown. Through a series of almost accidental experimental events, it was discovered that a chicken who had had its bursa surgically removed couldn't make antibodies. This was the pivotal discovery from which modern immunology and its practical applications—immune therapies by monoclonal antibodies and cytokines or lymphokines, organ transplantation, and the promised great immune therapies—devolve.

That's the way it is; there is a large element of inefficiency, an unpredictability in the biomedical sciences. Without those "good-for-nothing"-ologies many, if not most, of the great-leap-forward discoveries benefiting our health and well-being would not have had their seed origins. The foundations and government agencies that support biomedical science must somehow be made to realize that contrary to the laws of logic and economics, all natural science pursuits must be sustained. It is just as important, in the long run, to train and support the person whose life is devoted to the study of the bumblebee as it is to train and support the person whose life is devoted to the study of the AIDS virus. Those whose curiosity leads to the taxonomy of mosses, the physiology of the firefly, the tapeworms of sharks are our great resource.

Their curiosity is humanity's good fortune. From somewhere in the whole pool of knowledge the new advances will emerge. If all the sciences are not encouraged, the disciplines that effect our health and well-being will ultimately wither at a dead end.

Index